Marine Pollution - Recent Developments

Edited by Monique Mancuso, Mohamed H.H. Abbas, Teresa Bottari and Ahmed A. Abdelhafez

Published in London, United Kingdom

Marine Pollution - Recent Developments
http://dx.doi.org/10.5772/intechopen.100945
Edited by Monique Mancuso, Mohamed H.H. Abbas, Teresa Bottari and Ahmed A. Abdelhafez

Contributors
Md. Wahidul Alam, Pathmalal M. Manage, Bandara Kumudu R.V., Bilal Hammoud, Norbert Wehn, Chihhao Fan, Fatima Haque, Fikile Portia Ndlovu, Mads Nistrup Madsen, Henrik Skov, Michael Potthoff, Néstor Hernando Campos, José Luis Marrugo-Negrete, Andrew Tamale, Justine Okello, Celsus Sente, Ahmed A. Abdelhafez, Abdel Aziz Tantawy, Mohamed H.H. Abbas, Shawky M. Metwally, Amera Sh. Metwally, Aya Sh. Metwally, Rasha R.M. Mansour, Sedky H. Hassan, Hassan H. Abbas, Ihab M. Farid, Nermeen N. Nasralla, Ahmed S.H. Soliman, Mohammed E. Younis, Ghada S.A. Sayed, Mahfouz Z. Ahmed, Mahdy H. Hamed, Mahmoud I. El-Kelawy, Gamal Hassan Kamel, Hussein Ferweez, Ahmed M. Diab, Ehdaa Alaa Mohamed Abed, Ahmed Farouk Al-Hossainy, Heidi Ahmed Ali Abouzeid

First published in London, United Kingdom, 2023 by IntechOpen
IntechOpen is the global imprint of INTECHOPEN LIMITED, registered in England and Wales, registration number: 11086078, 5 Princes Gate Court, London, SW7 2QJ, United Kingdom
Printed in Croatia

British Library Cataloguing-in-Publication Data
A catalogue record for this book is available from the British Library

Additional hard and PDF copies can be obtained from orders@intechopen.com

Marine Pollution - Recent Developments
Edited by Monique Mancuso, Mohamed H.H. Abbas, Teresa Bottari and Ahmed A. Abdelhafez
p. cm.

This title is part of the Environmental Sciences Book Series, Volume 5
Topic: Pollution
Series Editor: J. Kevin Summers
Topic Editors: Ismail Md. Mofizur Rahman and Zinnat Ara Begum

Print ISBN 978-1-80356-299-5
Online ISBN 978-1-80356-300-8
eBook (PDF) ISBN 978-1-80356-301-5
ISSN 2754-6713

For EU product safety concerns:
IN TECH d.o.o., Prolaz Marije Krucifikse Kozulić 3, 51000 Rijeka, Croatia,
info@intechopen.com or visit our website at intechopen.com.

IntechOpen Book Series
Environmental Sciences
Volume 5

Aims and Scope of the Series

Scientists have long researched to understand the environment and man's place in it. The search for this knowledge grows in importance as rapid increases in population and economic development intensify humans' stresses on ecosystems. Fortunately, rapid increases in multiple scientific areas are advancing our understanding of environmental sciences. Breakthroughs in computing, molecular biology, ecology, and sustainability science are enhancing our ability to utilize environmental sciences to address real-world problems.

The four topics of this book series - Pollution; Environmental Resilience and Management; Ecosystems and Biodiversity; and Water Science - will address important areas of advancement in the environmental sciences. They will represent an excellent initial grouping of published works on these critical topics.

Meet the Series Editor

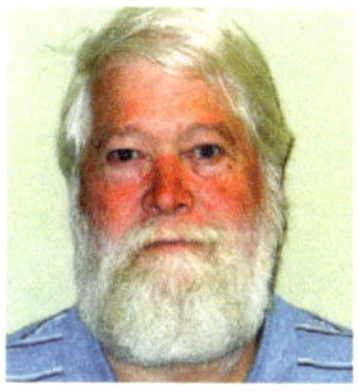

J. Kevin Summers is a Senior Research Ecologist at the Environmental Protection Agency's (EPA) Gulf Ecosystem Measurement and Modeling Division. He is currently working with colleagues in the Sustainable and Healthy Communities Program to develop an index of community resilience to natural hazards, an index of human well-being that can be linked to changes in the ecosystem, social and economic services, and a community sustainability tool for communities with populations under 40,000. He leads research efforts for indicator and indices development. Dr. Summers is a systems ecologist and began his career at the EPA in 1989 and has worked in various programs and capacities. This includes leading the National Coastal Assessment in collaboration with the Office of Water which culminated in the award-winning National Coastal Condition Report series (four volumes between 2001 and 2012), and which integrates water quality, sediment quality, habitat, and biological data to assess the ecosystem condition of the United States estuaries. He was acting National Program Director for Ecology for the EPA between 2004 and 2006. He has authored approximately 150 peer-reviewed journal articles, book chapters, and reports and has received many awards for technical accomplishments from the EPA and from outside of the agency. Dr. Summers holds a BA in Zoology and Psychology, an MA in Ecology, and Ph.D. in Systems Ecology/Biology.

Meet the Volume Editors

Monique Mancuso is a biologist with an MSc in Biological Sciences (1999) and a Ph.D. in Marine Environmental Health, Fish Pathology and Quality (2006), both from the University of Messina, Italy. She is also a specialist in applied microbiology. She has been a researcher at the Institute for Marine Biological Resources and Biotechnology of the National Research Council (IRBIM-CNR) of Messina since 2013. Dr. Mancuso is a member of the editorial board of several scientific journals. Her research interests include bacterial diseases in wild and farmed marine organisms, synthetic antibacterial effects, antibacterial effects of natural essential oils, and the evaluation and characterization of microplastics in seawater and marine organisms.

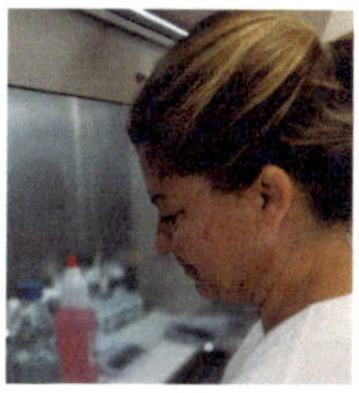

Teresa Bottari obtained a DVM and Ph.D. from the University of Messina, Italy. Since 2008, she has been a researcher at the Institute for Marine Biological Resources and Biotechnology of the National Research Council (IRBIM-CNR) of Messina. Dr. Bottari participates in several projects, including MEDITS (evaluation of demersal resources in the Mediterranean Sea), CAMPBIOL (monitoring of commercial catches), and RITMARE (use of fish parasites as biological tags for stock discrimination). Her research interests include the growth and reproduction of demersal and benthic resources, use of fish parasites as biological tags, and the impact of plastics on the marine environment (evaluation and characterization of plastics pollution levels in marine organisms).

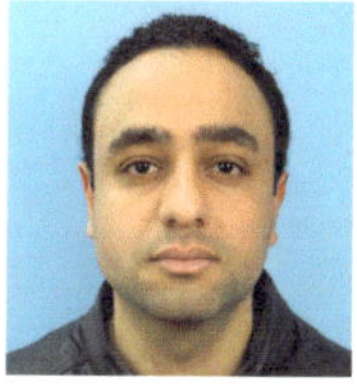

Ahmed A. Abdelhafez, Ph.D., is a professor and head of the Department of Soils and Water Science, Faculty of Agriculture, New Valley University, Egypt. He is one of the leading scientists in the field of biochar in the Arab region. He worked as a researcher at the Department of Environmental Research, Agricultural Research Center (ARC), Egypt, for more than 10 years. Dr. Abdelhafez focuses mainly on agricultural production, environmental contamination control, risk assessment, and biochar technology. He is a member of the National Committee of Soil Sciences and the Academy of Scientific Research & Technology, Egypt. He has published several research papers on environmental contamination, risk assessment, and potential remediation technologies. Dr. Abdelhafez is a member of the editorial board of several scientific journals and the chief editor of *Biochar and Compost Technology.*

Mohamed H.H. Abbas, Ph.D., is a Professor of Soil Chemistry, Faculty of Agriculture, Benha University, Egypt. He began his academic career as a demonstrator at Zagazig University, Egypt, in 2000. He received his MA and Ph.D. from Benha University. He is a board member of the *Egyptian Journal of Soil Science.* His research focuses on the sorption/desorption of potentially toxic elements in soil, their environmental pathways, related risk assessment, and environmental contamination control. He also conducts research on biochar, land reclamation, and the chemistry of organic matter in soil and plant nutrition. Dr. Abbas is a reviewer for several national and international journals.

Contents

Preface

Unfortunately, today the seas and oceans are increasingly polluted due to human activities and although much has been done to address this huge problem, the steps that must be taken to solve it are numerous and difficult.

This book provides a comprehensive overview of marine pollution. It includes 10 chapters that deal with various topics related to marine pollution, ranging from microplastics to the dispersion of oil in the sea to contamination by potentially toxic elements (PTEs) such as mercury (Hg). All the chapters emphasize the need for universally adopted policies that use common mitigation strategies to effectively address the problem of marine pollution in a unique way.

Monique Mancuso and Teresa Bottari
Institute for Biological Resources and Marine Biotechnology (IRBIM),
CNR - Messina, Italy

Ahmed A. Abdelhafez
Professor,
Faculty of Agriculture,
Department of Soils and Water Science,
New Valley University,
New Valley, Egypt

Mohamed H.H. Abbas
Professor,
Faculty of Agriculture,
Benha University,
Benha, Egypt

Section 1

Marine Pollution with Inorganic Pollutants

Chapter 1

Heavy Metal Contamination in the Coastal Environment and Trace Level Identification

Bandara Kumudu R.V. and Pathmalal M. Manage

Abstract

Heavy metal pollution in the coastal environment is a great concern as its adverse effects on marine health. Heavy metals are a group of persistent organic pollutants and last for years in the environment. Due to their widespread distribution, high hydrophobicity, prolonged persistence, and negative effects on the environment and human health, these chemicals have raised attention. Due to a lack of research and advanced detection techniques, heavy metal pollution in coastal areas of some Asian countries is critical. Scientists have developed several methods for detecting heavy metals in the environment, including atomic absorption spectroscopy, inductively coupled plasma massspectroscopy, and high-performance liquid chromatography coupled with electrochemical or UV-Vis-detectors. However, a newly optimized, sensitive, cost-effective, and precise technology for detecting heavy metals at ultra-trace levels is solid phase micro-extraction and gas chromatography mass-spectrometry. As a result, the book chapter will describe the theoretical, practical approach, and modern technology for detecting and quantifying heavy metal contaminations in the marine ecosystem, including the effects of heavy metals on the marine animals, human and environmental health, and challenges and future perspectives of heavy metal degradation using a green approach, as well, the effects of heavy metals on the marine animal, human, and environmental health.

Keywords: heavy metals, pollutant, marine ecosystem, SPME, trace level

1. Introduction

Chemical production has increased from one million tons in 1930 to over 400 million tons now, with around 100,000 different chemical substances registered on the European Union market, of which 30,000 to 70,000 are used daily [1, 2]. The most of chemicals in water bodies, including heavy metals, derive from wastewater from industrial, agricultural, and domestic sources, as well as municipal sewage treatment plants [3]. Around 10% of the globally available runoff is being used by industries and municipalities, resulting in a stream of effluent that flows into rivers, lakes, groundwater, or coastal water. As a result, each year over 300 million tons of heavy metals from industrial and consumer products, including Cr, Cu, Zn, As, Cd, Pb, and Sn, find their way into natural waters [4]. Cement plants, coal and energy bases, textiles, ship breaking/recycling, and tanneries are only a few examples of key types of businesses that have become important for macro and microeconomic growth as well as heavy metal pollution. Agriculture, which uses 140 million tons of fertilizers and several million tons of pesticides each year, contributes to further pollution [4]. Chemical pollution is mainly a threat to developing countries due to the lack of compliance with environmental and safety regulations. Low labor costs and favorable geographic locations have made the ship-repairing industry highly profitable, releasing considerable quantities of heavy metals into the environment [5]. Even at low levels of exposure, these metals are hazardous to the physical and chemical functions of animals and can damage their several organs such as brain, liver, and reproductive organs. As a result, monitoring and assessment can aid with coastal environmental protection management and planning. When using heavy metals, trace metals analysis can be used to detect and identify small amounts of metals in a sample, which is important for quality control and regulatory compliance.

When using heavy metals, trace metals analysis can be used to detect and identify small amounts of metals in a sample, which is important for quality control and regulatory compliance.

1.1 Heavy metals

In comparison to water, heavy metals are metallic elements with a higher density. Assuming that toxicity and heaviness are connected. Heavy metals also include metalloids such as arsenic, tin, lead, etc., which can cause harmful effects at low concentrations [6]. Environmental contamination of these metals is responsible for threatening the environment and global public health. Due to vast development in their use in numerous industrial, agricultural, domestic, and technological applications, human exposure has increased. Heavy metals in the environment could be found in geological, pharmaceutical, industrial, agricultural, atmospheric, and domestic effluent sources (**Table 1**). Smelters, foundries, mining, and other metal-based industrial operations are all major contributors to environmental contamination [7].

Heavy metals are naturally occurring elements found throughout the earth's crust; however, human activities such as mining and smelting, industrial application, and

Heavy metals	Application
Nickel (Ni)	Batteries processing units, Galvanization, metal refining, fertilizers, painting industries
Zinc (Zn)	Painting and dye, wood preservatives, fertilizers, rubber and detergent industries
Arsenic (As)	Wood preservatives, dye and automobile industries
Cadmium (Cd)	Painting, galvanized, pesticide, polyvinyl plastic, refined petroleum products industries
Aluminium (Al)	Aluminium phosphate Pesticide, ceramics, automotive parts industries
Iron (Fe)	Engine parts and metal refining industries
Copper (Cu)	Electroplating, metal refining and plastic industries
Lead (Pb)	Pesticides, mobile batteries, petrol based materials and leaded gasoline industries
Chromium (Cr)	Electroplating, leather, textile, pulp, tanning, and chrome plating industries
Tin (Sn)	Wood preservatives, Leather, Textile dye, antifouling paint, pesticide industries

Table 1.
Industrial uses of the heavy metals.

domestic and agricultural use of metals and metal-containing compounds end up causing the vast majority of environmental contamination and human exposure. Heavy metals are usually found in trace amounts in the earth's crust, ranging from a few parts per trillion (ppt) for noble metals to up to 5% for iron. They could be encountered in their elemental, metallic form. Weathering and erosion led them to leach into the soil, rivers, and groundwater. When the Earth's mantle was still liquid 4–5 billion years ago, heavy metals sank to the center and formed the iron- and nickel-rich core [8].

Due to their trace concentrations (ppt range to less than 10ppm) in a wide range of environmental matrices, heavy metals are classified as trace elements. Physical factors such as phase association, temperature, sequestration, and adsorption affect heavy metal bioavailability. Chemical parameters, such as lipid solubility, complexation kinetics, and octanol/water partition coefficients, have an impact on speciation at thermodynamic equilibrium. Biological factors include species characteristics, biochemical/physiological adaptability, and trophic interactions.

1.2 The significance of heavy metals to animal and plant life

Metals are required for the proper performance of several biochemical and physiological processes in humans, animals, and plants. Microelements are trace elements having minor dietary requirements, such as chromium (Cr), iron (Fe), cobalt (Co), manganese (Mn), copper (Cu), molybdenum (Mo), zinc (Zn), and selenium (Se). They are found in trace levels (ppt, ppb, or ppm) in a range of matrices and their bioavailability varies [9]. Trace elements are commonly added to animal feed as a nutritional supplement to enhance health and productivity. Heavy metals have been shown to impact a variety of cellular organelles and components, including the cell membrane, mitochondria, lysosome, endoplasmic reticulum, nuclei and enzymes involved in metabolism, detoxification, and damage repair. Excessive exposure to these elements at elevated concentrations has been associated with cellular or systemic problems and may be a source of pollution [10].

Chlorophyll production, protein modifications, DNA synthesis, photosynthesis, redox reactions in the chloroplast and mitochondria, nitrogen fixation, and sugar metabolism are all affected by heavy metals. More than 300 enzymes and 200 transcription factors required zinc as a cofactor for membrane integrity maintenance, auxin metabolism, and reproduction [11]. However, the remaining excess of heavy metals in the environment is responsible for harmful consequences.

2. Increasement of heavy metals in the coastal environment

Metal corrosion, air deposition, soil erosion of metal ions and heavy metal leaching, sediment resuspension, and metal evaporation from water resources to soil and groundwater are all potential sources of environmental contamination. Natural occurrences such as volcanic eruptions and weathering are also identified as major contributors to heavy metal contamination. Metal refineries, petroleum combustion, high-tension lines, coal combustion in power plants, nuclear power stations, microelectronics, wood preservation, plastics, textiles, and paper processing industries have all been industrial sources [7]. Tin (Sn), molybdenum (Mo), cobalt (Co), copper (Cu), selenium (Se), zinc (Zn), chromium (Cr), nickel (Ni), magnesium (Mg), manganese (Mn), and iron (Fe) are reported to be essential nutrients needed for a variety of physiological and metabolic activities. Deficiencies in these micronutrients can lead to a range of syndromes and disorders.

Land-based metal polluting industries, such as textiles, coal and energy bases, cement plant, leather, and ship breaking/recycling, have expanded significantly over the past few decades as these types of enterprises have become an important factor for macro and micro perspectives of economic growth. These industries release significant amounts of heavy metals into the environment indicating opportunities and possible risks to a more globalized economy. However, about 40% of the world's seas have been significantly impacted by human activity, with the most severe consequences occurring in coastal regions [12, 13]. Indeed, anthropogenic metal contamination in coastal and marine environments has been a major issue since it may have toxic effects on aquatic living organisms, destroy natural ecosystems, and significant health risks to humans through consumption of contaminated seafood [14].

2.1 Heavy metal contamination in the marine ecosystem

Metal concentrations in the marine environment are estimated to be in the nanogram to microgram per liter (liquid phase) or per gram range as they are found naturally components of the earth's crust (solid phase). Concentrations of important heavy metals (Zn, Pb, Cd, Cu, Sn, and Hg) in the marine environment have increased by five to ten times in the past few decades compared to values recorded fifty to one hundred years ago [15]. The global increase in metal contamination in marine ecosystems has been mainly driven by economic development and accelerated industrialization in recent decades. Metals enter aquatic systems and accumulate in various ways, as shown in **Figure 1**. Changes in physicochemical parameters such as salinity, redox potential, temperature, pH, and organic ligand concentrations could cause metals

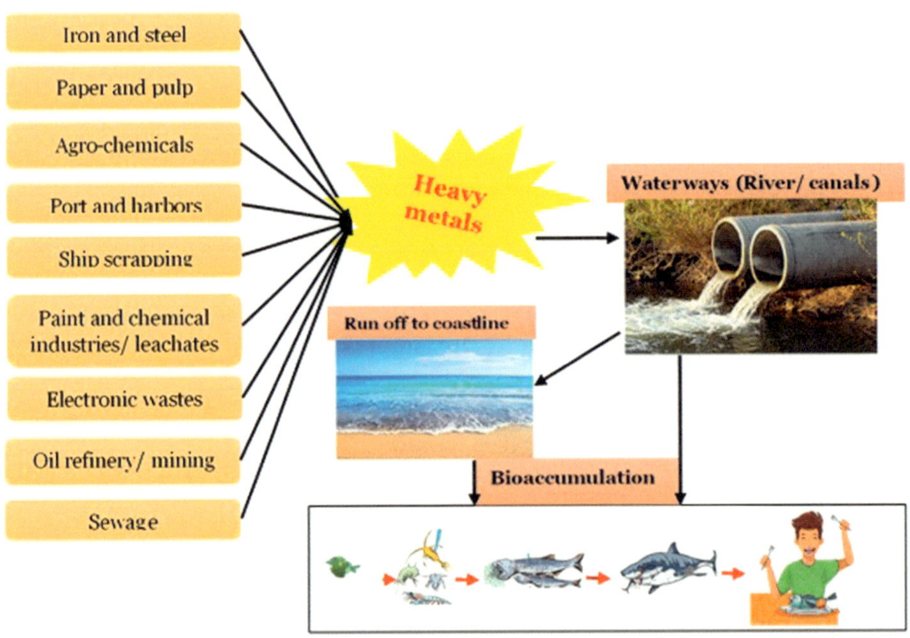

Figure 1.
Major routes for metals contamination and bioaccumulation in the aquatic environment.

to disintegrate from a solid phase once adsorbed into marine sediments. As a consequence, the toxicity, mobility, and bioavailability of metals are largely determined by the environment [15, 16]. Important terrestrial and aquatic ecosystems, including mangrove forests, intermittently and permanently flooded wetlands, and tidal flats, are found in the coastal environment, which is highly dynamic and ecologically complex. These enriched ecosystems are at risk due to environmental consequences and human encroachment. The wastes from scrapped ships, including oils and persistent organic pollutants (POPs), enter the coastal bay, which is home to a variety of marine life, including many endangered and vulnerable species; pollution has been identified as one of three major transboundary issues affecting the marine ecosystem. Heavy metals can accumulate in aquatic habitats to toxic levels because they are not biodegradable, although some metals are essential to the ecosystem's function. Metals are eventually integrated into the sediment at the bottom, where organisms (benthic organisms) can gather them. Heavy metals affect aquatic life at greater levels. Cadmium and copper concentrations in the water surrounding shipbreaking zones were high enough to affect fish spawning and death. In coastal areas, a higher concentration of mercury and tin has delayed the development of mollusks, and lead has reduced the nesting capacity of marine birds [17].

Low levels of dissolved metals in seawater may or may not present a hazard to marine biota but some species have shown a tendency to accumulate the metals from the water. Li et al. [18] studied the effects of chelating agents on the uptake and accumulation of cadmium by *Mytilus edulis*. Complexation of cadmium with either EDTA, humic and alginic acids, or pectin doubles both rate of accumulation and the final tissue concentration. Yap et al. [19] determined trace metal (Hg, Cd, Cu, Ag, Zn, and Pb) concentrations in mussels from the estuaries were compared with those of coastal and offshore regions in the southeastern North Sea. Bandara et al. [20] reported that tributyltin contamination in marine gastropod; *Thais clavigera* and mussels, *Perana perana*, *Perna viridis*, *Crassostrea madrasensis,* and *Crassostrea cuculata*. Levels of selected heavy metals in sediments collected from subtidal areas off the Iranian coastline of the Arabian Gulf were measured by Pourang et al. [21]. This study concluded that the concentrations of Cd, Pb, and Ni in the sediments (2.9, 90.5, and 64.9 μg/ g dry weight, respectively) were notably higher than global baseline values. Assessment of contaminants in Dubai coastal region, United Arab Emirates, was conducted by Al-Darwish et al. [22]. This study found that the highest concentrations of the selected metals were found in chronically polluted areas and reported elevated levels of Cu, Ni, and Zn in comparison with background levels of unpolluted sediments in the Arabian Gulf, which is heavily occupied by a variety of industries.

3. Toxicity of heavy metals

Heavy metals: Fe, Co, Cu, Mn, Mo, Se, Zn, Cr, and Cd, as well as Hg, Pb, and As, have a high density compared to water and are present in trace amounts in different matrices. Heavy metals could cause toxicity at low concentrations, hence their heaviness and toxicity are associated [23]. Heavy metals have a negative impact on the soil, water, air, plants, animals, and humans. High heavy metal concentrations in soil can impact soil quality, particularly pH, color, porosity, and natural composition, as well as crop yield and the extinction of many species of normal flora and animals. Their buildup in the water creates significant issues for humans and ecosystems due to a decrease in drinking water quality and cleanliness, as well as a decrease in water supplies for all life forms. Increased acid rain, corrosion, eutrophication and haze, decreased agricultural yields, and a shortage of oxygen can be caused by heavy metal contamination in the air. They can damage plants' roots and leaves, disrupt important biochemical processes such as photosynthesis, impair mineral absorption, damage chlorophyll, and inhibit root growth and development, leading to a reduction in total plant growth [24].

Heavy metal toxicity in animals shows itself with reduced body weight, renal damage, liver damage, shorter life span, increased oxidative stress, changes in cell composition, and DNA damage. In humans, they can cause renal illness, liver disease, lung complications, and some cancers [25]. Heavy metals accumulate in organs and soft tissues when they are not metabolized by the body, causing them toxic. Ingestion of contaminated food or water, inhalation, or skin absorption have all been routes they enter the human body. One of the most common ways for heavy metals to enter the bodies of animals is via ingestion. These metals can have inhibitory, stimulatory, or toxic effects on specific biochemical processes, resulting in a variety of health problems in the nervous system (Alzheimer's, Parkinson's, depression, and dementia), the bone system (bone mineralization), and the reproductive system. reactive oxygen species (ROS) can also cause DNA and RNA damage and malignancies of the lungs, skin, and bladder, as a consequence of its production. The amount of exposure, the length of exposure, the pollutant concentration, and the organisms exposed to it, as well as the metal's type and oxidation state, determine its toxicity [26].

3.1 Heavy metal toxicity amplification in plants and animals

Heavy metals persist in nature, posing a threat to human health in addition to their detrimental impacts on plants and wildlife. For example, lead (Pb) is one of the most toxic heavy metals, with a soil retention time of 150–5000 years and a reported concentration retention period of up to 150 years [27]. Plants that grow in heavy metal-contaminated areas tend to collect higher concentrations of heavy metals in the food chain. The main route for heavy metals into animal and human tissues would be through contaminated food, making them vulnerable to a variety of diseases.

Several heavy metals such as Cr, Cd, Pb, Hg, and Al are highly toxic even in trace concentrations, even though they are nonessential and have no physiological role [28]. Both essential and nonessential heavy metals damage plants in comparable ways, caused by low biomass accumulation, chlorosis, growth and photosynthetic inhibition, altered water balance and nutrient assimilation, and senescence, which leads to plant death.

3.2 Potential Human Exposure to heavy metals

Each year, over 300,000 employees are exposed to heavy metals and metal-containing compounds in the workplace. One of the biggest sources of health concern is industrial worker exposure to the high risk of Cr-induced diseases. Cr is anticipated to be released into the environment at a rate of 33 tons per year, which poses a significant carcinogenic risk. The Occupational Safety and Health Administration (OSHA) in the United States recently established a "safe" level of $5 g/m^3$ for an eight-hour time-weighted average. The overall human population's atmospheric levels range from 1 to $100 ng/cm^3$, while levels near Cr-related industries can exceed this range [10].

Nonoccupational exposure occurs when heavy metal-containing meals and water are consumed. Whenever it comes to chromium contamination, recorded levels range from 1 to 3000 mg/kg in soil, 5 to 800 g/L in seawater, and 26 g/L to 5.2 mg/L in rivers and lakes. The amount of chromium in food varies wildly depending on how it is processed and cooked. Fresh foods contain chromium levels ranging from 10 to 1,300 g/kg [29]. Employees in chromium-related industries can be exposed to two orders of magnitude higher levels of chromium than the general population.

Human exposure to heavy metals seems to be mostly by inhalation, with the lung being the primary target organ. However, significant human exposure to heavy metals has also been recorded through the skin [30]. For example, exposure to chromium found in cement is implicated in the worldwide incidence of dermatitis among construction workers. Exposure to Cr(VI)-containing chemicals in the workplace and the environment has been linked to multi-organ toxicity in humans, including kidney impairment, allergy and asthma, and lung cancer.

Enormous concentrations of heavy metals in the air would irritate the nose lining and lead to ulcers. The numerous health problems documented in animals after swallowing heavy metals at detectable levels include stomach irritation and ulcers, sperm loss, anemia, and endocrine abnormalities [8]. Some people who are extremely sensitive to heavy metals have experienced allergic reactions such as edema and severe redness of the skin. In both humans and animals, toxic substances in drinking water have been related to an increased risk of stomach cancer. Ingestion of extremely high concentrations of these compounds by humans, whether it be by accident or on purpose, has resulted in severe respiratory, cardiovascular, gastrointestinal, hematological, hepatic, renal, and neurological consequences, in people who died or survived as a result of medical care. Strong evidence of heavy metal carcinogenicity in humans and terrestrial mammals has also been reported by Mahurpawar [25] and Tchounwou et al. [10].

3.3 Heavy metal toxicity and carcinogenicity

Heavy metal toxicity is mostly determined by their oxidation state and lipophilicity. Even though the processes of biological interaction are unknown, the chemicals can penetrate through cell membranes, and their intracellular reduction to reactive intermediates may be connected to toxicity variation. It can be absorbed to some extent through the lungs, gastrointestinal tract, and even intact skin. It is considered a detoxification process when harmful substances are reduced at a distance from the target site for toxic or genotoxic action; nevertheless, when the compound is reduced in or near the cell nucleus of target organs, it may help activate their toxicity [31].

Under physiological conditions, heavy metals can be reduced by hydrogen peroxide (H_2O_2), ascorbic acid, and glutathione reductase to form reactive intermediates such as thiol radicals and hydroxyl radicals. Any of these heavy metal species could disrupt cellular integrity and function by attacking DNA, proteins, or membrane lipids [32].

With animal research, many harmful effects of heavy metals on mammals have been reported [25, 33]. After subcutaneous treatment of Cr, rats developed severe urea nitrogen, creatinine elevations, proteinuria, an increase in blood alanine aminotransferase activity, and hepatic lipid peroxide production. In similar studies, Sahu et al. [34] found that chromium induced kidney impairment in rats when administered as a single subcutaneous injection. Giving Cr to rats in water caused hepatic

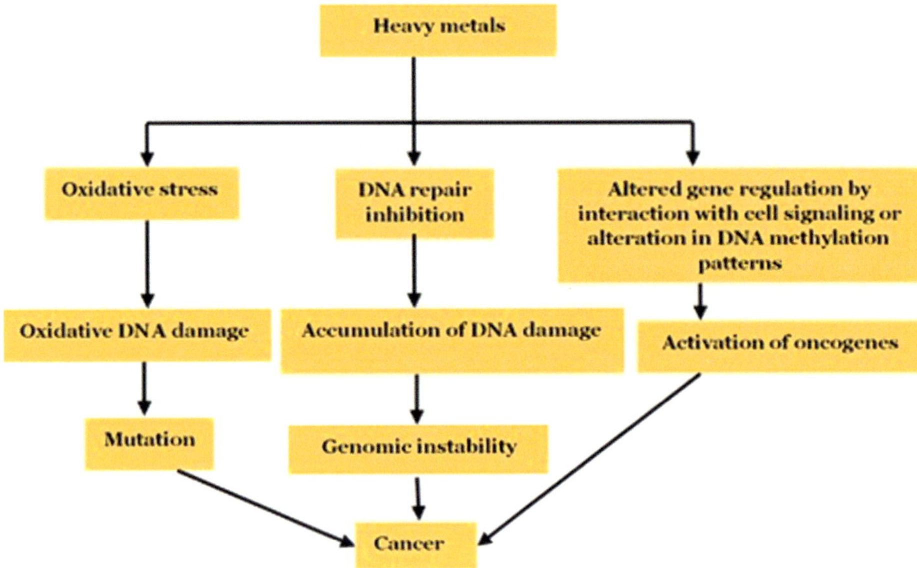

Figure 2.
Mechanism of heavy metal induced carcinogenicity (Source: Hartwig [37]).

mitochondrial and microsomal lipid peroxidation, as well as increased urine lipid metabolite excretion, including malondialdehyde.

In humans, heavy metals have been linked to harmful health consequences. Respiratory malignancies have been recorded among workers exposed to Cr, Sn, Pb, and As-containing substances in the workplace following the epidemiological research [35]. DNA strand breakage in peripheral lymphocytes and lipid peroxidation products in urine in exposed employees reinforce the proof of heavy metal-induced human toxicity. Oxidative damage is assumed to be the cause of genotoxic impacts, such as chromosomal abnormalities and DNA strand breaks. According to the current study by Singh et al. [36], non-oxidative processes appear to have a role in carcinogenesis. Carcinogenicity appears to be linked to inhalation of less soluble chemicals such as Sn and Cr. Surface charge, phagocytization capacity, crystal modification, and size are all features of chromium that may be significant in predicting cancer risk, according to epidemiological studies (**Figure 2**).

4. Heavy metals in the environment

Environmental research on heavy metals requires a basis in environmental chemistry, ecotoxicology, and ecology. The study of xenobiotics in food chains is a significant field of research with important environmental, ecological, and economic implications. According to Ali et al. [38], marine chemistry has public health implications: "Aquatic chemistry is a key component of public health".

Understanding heavy metal pollution in coastal areas of certain countries, such as Asian and Oceania countries, is vital due to a lack of research and advanced xenobiotic monitoring methods. In recent years, several methods for detecting heavy metals in environmental matrices have been developed. Supercritical fluid extractions [39], acid extraction, alkaline extraction digestion with sodium hydroxide, and subsequent extraction with tetra ethyl ammonium hydroxide are examples of these methods [40]. When heavy metals are present in matrices at very low levels (ppb and ppt), however, extracting these chemicals is time-consuming, costly, and a source of inaccuracy and cross-contamination [41]. As a result, in many regions around the world, detecting heavy metals in aquatic settings has become difficult and inaccessible. solid phase extraction (SPE) and liquid-liquid extraction (LLE) are the most prevalent methods for separating metal compounds from matrices [42]. SPE cartridges, on the other hand, are costly and nonreusable, while LLE extraction procedures require a substantial volume of organic solvents. Bandara et al. [20] have optimized an automated solid phase micro extraction (SPME) method to detect tributyltin at ultra-trace concentrations (ppt levels) because it reduces sample preparation, solvent usage, and analytical costs.

For the quantitative determination of metals, atomic absorption spectrometry (AAS), inductively coupled plasma-mass spectrometry (ICP-MS), and gas chromatography-mass spectrometry (GC-MS) are widely employed.

4.1 Heavy metal quantification

For the assessment and implementation of heavy metal pollution management methods, environmental monitoring and quantification are required. The concentrations of potentially toxic metals and metalloids in diverse environmental media, such as water, sediments, and biota, should be monitored regularly. This environmental analysis will provide important information on the distribution, major sources, destination, and bioaccumulation of these elements in the environment, as well as on their bioaccumulation.

4.2 Assessment of heavy metal pollution using biomarkers and bioindicators

Łuczynska et al. [43] explain the use of bioindicators for heavy metal pollution monitoring and evaluation as follows: "Measuring metal concentrations in selected species of the resident biota could provide a more meaningful assessment of the impact of metal pollution." To assess heavy metal contamination and environmental pollution, a variety of plant and animal species have been used as bioindicators. Bandara et al. [20] studied the impact of tributyltin contamination in the Sri Lankan coastal stretch employing mollusks *Perna perna*, *Thias clavigera*, and *Perna viridis* at commercial and fishery harbors as the first record of tributyltin contamination.

4.3 Atomic absorption spectrometry (AAS)

By its simplicity and the fact that it can quantify a large number of metals (cobalt, chromium, cadmium, copper, iron, manganese, nickel, lead, and zinc) with a minimum detection limit of 1 ppb, flame atomic absorption spectrometry (FAAS) is extensively used for metal determination from the soil, water, and biological samples. A graphite furnace and greater atomization temperatures are required for electrothermal atomic absorption spectrometry (ETAAS). This method has the advantage of requiring a minimal sample volume (20–50 uL) and having excellent minimum detection limits at parts per billion level.

The use of chemical modifiers in ETAAS analysis of volatile elements (arsenic, antimony) is required to stabilize the analysis, which would otherwise evaporate at temperatures above 400°C. One of the most successful methods for determining trace elements in various matrices is chemical vapor production in combination with atomic absorption spectroscopy. HGAAS (hydride generation atomic absorption spectroscopy) and cold vapor atomic absorption spectroscopy are two examples of this technology (CVAAS). HGAAS is used for the analysis of hydride-forming metals (selenium, arsenic, tin, and lead) while CVAAS is used for mercury analysis from various samples.

4.4 Mass spectrometry with inductively coupled plasma (ICP-MS)

ICP-MS (inductively coupled plasma-mass spectrometry) is a method for analyzing multi elements at trace levels in a range of liquid samples. It has a high level of precision and accuracy. As a result, solid materials must be digested before the examination, and digestion methods require the use of HNO_3 or HNO_3 and H_2O_2. In argon plasma, the material is atomized and ionized, and the resulting ions are sorted by a mass analyzer according to their mass-charge ratio (m/z) before being detected at the detector. The detection limits of this method are 0.2 ppt in solution and 0.1 ppb in tissues.

4.5 Solid phase micro extraction (SPME) and GCMS method

SPME is a solvent-free method that was developed by Pawliszyn in 1989 and combines sample extraction and concentration in a single process [44]. It consists of an organic phase-coated fiber that uses absorptive/adsorptive processes to selectively extract and concentrate the analytes present. SPME- GCMS procedure has mainly four steps that include extraction, derivatization, separation, and final detection. Ion exchange and gas chromatographic (GC) systems have been used to detect metal compounds directly from aquatic samples. The advantages of heavy metal determination include the availability of compound-sensitive, simple, cost-effective, and selective detectors (MS) with high-resolution power methods [45]. The recovery rate

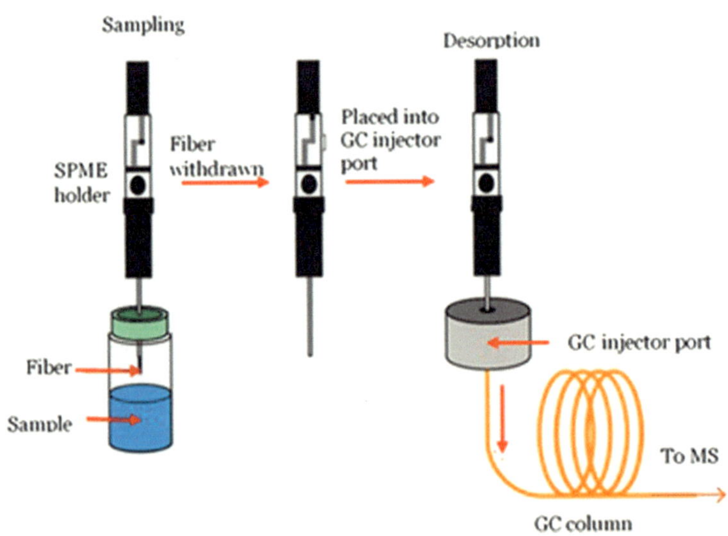

Figure 3.
SPME-GCMS analysis protocol of TBT. Source: Schmidt and Podmore, [46].

of this method to detect the quantity of organotin (Sn) in marine water was found to be 88% by Bandara et al. [20].

Heavy metal usage and contamination of the environment could come from active maritime heavy boating and shipping activities in coastal areas, as well as other industrial activities. As a result, SPME is a cost-effective, precise, and sensitive standardized method to detect and quantify heavy metals at the parts per trillion level anywhere in the world (**Figure 3**).

5. Conclusions

In both aquatic and terrestrial environments, heavy metals and metalloids are common pollutants. The hazard of an environmental chemical is determined by its persistence in the environment, as well as its toxicity and bioaccumulative potential. Toxic environmental pollutants that are persistent and bioaccumulative are more hazardous. persistence, bioaccumulation, and toxicity (PBT) are the three characteristics of heavy metals that make them hazardous. Some of the most environmentally hazardous heavy metals and metalloids are Cr, Ni, Cu, Zn, Cd, Pb, Hg, and As. These components' trophic transmission in aquatic and terrestrial food chains/webs has significant implications for animal and human health. The concentrations of potentially hazardous heavy metals and metalloids in various environmental segments as well as in the resident biota must be measured and monitored. Environmental chemistry and ecotoxicology of hazardous heavy metals and metalloids reveal that measures should be done to decrease their impact on human health and the environment. Some suggestions are as follows:

I. Heavy metal and metalloid background concentrations should be documented in various environmental media around the world for future reference.

II. Assess and record the levels of potentially toxic heavy metals and metalloids in water, sediments, and resident biota regularly.

III. Regular studies should be conducted to quantify the daily consumption of freshwater fish and other consumables, such as rice, by the world's resident population. This data will aid in a more reliable and accurate evaluation of human and environmental risk.

IV. Heavy metal contamination in aquatic and terrestrial ecosystems should be kept to a minimum to protect the biota and the health of its consumers.

V. The adverse impacts of toxic heavy metals on human health and the environment should be made widely understood.

VI. Before being released into natural waterways, industrial wastewaters must be properly treated.

VII. A priority for protecting human health and the environment should be encouraging and promoting scientific research on the environmental assessment of harmful substances, particularly heavy metals and metalloids.

Author details

Bandara Kumudu R.V.[1,2] and Pathmalal M. Manage[1,2*]

1 Centre for Water Quality and Algae Research, Department of Zoology, University of Sri Jayewardenepura, Sri Lanka

2 Faculty of Graduate Studies, University of Sri Jayewardenepura, Sri Lanka

*Address all correspondence to: pathmalal@sjp.ac.lk

References

[1] Abelkop AD, Graham JD. Regulation of chemical risks: Lessons for reform of the toxic substances control act from Canada and the European Union. Pace Environmental Law Review. 2015;**32**:108

[2] Trasande L, Zoeller RT, Hass U, Kortenkamp A, Grandjean P, Myers JP, et al. Estimating burden and disease costs of exposure to endocrine-disrupting chemicals in the European Union. The Journal of Clinical Endocrinology & Metabolism. 2015;**100**(4):1245-1255

[3] Patel P, Raju NJ, Reddy BC, Suresh U, Sankar DB, Reddy TVK. Heavy metal contamination in river water and sediments of the Swarnamukhi River Basin, India: Risk assessment and environmental implications. Environmental Geochemistry and Health. 2018;**40**(2):609-623

[4] Ochieng EZ, Lalah JO, Wandiga SO. Analysis of heavy metals in water and surface sediment in five rift valley lakes in Kenya for assessment of recent increase in anthropogenic activities. Bulletin of Environmental Contamination and Toxicology. 2007;**79**(5):570-576

[5] Sarraf, M., Stuer-Lauridsen, F., Dyoulgerov, M., Bloch, R., Wingfield, S. and Watkinson, R., 2010. The ship breaking and recycling industry in Bangladesh and Pakistan. Report No 58275-SAS.

[6] Alloway BJ. Heavy Metals in Soils: Trace Metals and Metalloids in Soils and Their Bioavailability. Dordrecht, The Nertherlands: Springer Science & Business Media; 2012

[7] Nagajyoti PC, Lee KD, Sreekanth TVM. Heavy metals, occurrence and toxicity for plants: A review. Environmental Chemistry Letters. 2010;**8**(3):199-216

[8] Engwa GA, Ferdinand PU, Nwalo FN, Unachukwu MN. Mechanism and health effects of heavy metal toxicity in humans. In: Poisoning in the Modern World-new Tricks for an Old Dog; Intechopen. 2019

[9] Gondal AH, Zafar A, Zainab D, Toor MD, Sohail S, Ameen S, et al. A detailed review study of zinc involvement in animal, plant and human nutrition. Indian Journal of Pure & Applied Biosciences. 2021;**9**(2):262-271

[10] Tchounwou PB, Yedjou CG, Patlolla AK, Sutton DJ. Heavy metal toxicity and the environment. Molecular, Clinical and Environmental Toxicology. 2012;**6**:133-164

[11] Dutta S, Mitra M, Agarwal P, Mahapatra K, De S, Sett U, et al. Oxidative and genotoxic damages in plants in response to heavy metal stress and maintenance of genome stability. Plant Signaling & Behavior. 2018;**13**:e1460048

[12] Lotze HK, Lenihan HS, Bourque BJ, Bradbury RH, Cooke RG, Kay MC, et al. Depletion, degradation, and recovery potential of estuaries and coastal seas. Science. 2006;**312**(5781):1806-1809

[13] Halpern BS, Walbridge S, Selkoe KA, Kappel CV, Micheli F, d'Agrosa C, et al. A global map of human impact on marine ecosystems. Science 2008;**319**(5865):948-952

[14] Naser HA. Assessment and management of heavy metal pollution in the marine environment of the Arabian Gulf: A review. Marine Pollution Bulletin 2013;**72**(1):6-13.

[15] Mashiatullah A, Chaudhary MZ, Ahmad N, Javed T, Ghaffar A. Metal pollution and ecological risk assessment in marine sediments of Karachi Coast, Pakistan. Environmental Monitoring and Assessment. 2013;**185**(2):1555-1565

[16] Sany SBT, Salleh A, Sulaiman AH, Sasekumar A, Rezayi M, Tehrani GM. Heavy metal contamination in water and sediment of the Port Klang coastal area, Selangor, Malaysia. Environmental Earth Sciences. 2013;**69**(6):2013-2025

[17] Kibria G, Hossain MM, Mallick D, Lau TC, Wu R. Trace/heavy metal pollution monitoring in estuary and coastal area of Bay of Bengal, Bangladesh and implicated impacts. Marine Pollution Bulletin. 2016;**105**(1):393-402

[18] Li J, Chapman EC, Shi H, Rotchell JM. PVC does not influence cadmium uptake or effects in the mussel (Mytilus edulis). Bulletin of Environmental Contamination and Toxicology. 2020;**104**(3):315-320

[19] Yap CK, Sharifinia M, Cheng WH, Al-Shami SA, Wong KW, Al-Mutairi KA. A commentary on the use of bivalve mollusks in monitoring metal pollution levels. International Journal of Environmental Research and Public Health. 2021;**18**(7):3386

[20] Bandara KRV, Chinthaka SDM, Yasawardene SG, Manage PM. Modified, optimized method of determination of Tributyltin (TBT) contamination in coastal water, sediment and biota in Sri Lanka. Marine Pollution Bulletin. 2021;**166**:112202

[21] Pourang N, Nikouyan A, Dennis JH. Trace element concentrations in fish, surficial sediments and water from northern part of the Persian Gulf. Environmental Monitoring and Assessment. 2005;**109**(1):293-316

[22] Al Darwish HA, Abd El-Gawad EA, Mohammed FH, Lotfy MM. Assessment of organic pollutants in the offshore sediments of Dubai, United Arab Emirates. Environmental Geology. 2005;**48**(4):531-542

[23] Nachana'a Timothy ETW. Environmental pollution by heavy metal: An overview. Chemistry. 2019;**3**(2):72-82

[24] Rehman AU, Nazir S, Irshad R, Tahir K, Rehman K, Islam RU, et al. Toxicity of heavy metals in plants and animals and their uptake by magnetic iron oxide nanoparticles. Journal of Molecular Liquids. 2021;**321**:114455

[25] Mahurpawar M. Effects of heavy metals on human health. International Journal of Reseacrh-Granthaalayah. 2015;**50**:2394-3629

[26] Carpenter DO. Effects of metals on the nervous system of humans and animals. International Journal of Occupational Medicine and Environmental Health. 2001;**14**(3):209-218

[27] Yang L, Parkin DM, Ferlay J, Li L, Chen Y. Estimates of cancer incidence in China for 2000 and projections for 2005. Cancer Epidemiology and Prevention Biomarkers. 2005;**14**(1):243-250

[28] Gill M. Heavy metal stress in plants: A review. International Journal of Advanced Research. 2014;**2**(6):1043-1055

[29] Zhang Y, Wang Y, Li M, Liu S, Yu J, Yan Z, et al. Traditional uses, bioactive constituents, biological functions, and safety properties of Oviductus ranae as functional foods in China. Oxidative Medicine and Cellular Longevity. 2019;**2019**:1-24

[30] Fu Z, Xi S. The effects of heavy metals on human metabolism.

Toxicology Mechanisms and Methods. 2020;**30**(3):167-176

[31] Kocadal K, Alkas FB, Battal D, Saygi S. Cellular pathologies and genotoxic effects arising secondary to heavy metal exposure: A review. Human & Experimental Toxicology. 2020;**39**(1):3-13

[32] Jaishankar M, Tseten T, Anbalagan N, Mathew BB, Beeregowda KN. Toxicity, mechanism and health effects of some heavy metals. Interdisciplinary Toxicology. 2014;7(2):60-75

[33] Ali H, Khan E. Trophic transfer, bioaccumulation, and biomagnification of non-essential hazardous heavy metals and metalloids in food chains/webs—Concepts and implications for wildlife and human health. Human and Ecological Risk Assessment: An International Journal. 2019;**25**(6):1353-1376

[34] Sahu BD, Koneru M, Bijargi SR, Kota A, Sistla R. Chromium-induced nephrotoxicity and ameliorative effect of carvedilol in rats: Involvement of oxidative stress, apoptosis and inflammation. Chemico-Biological Interactions. 2014;**223**:69-79

[35] Moreno T, Trechera P, Querol X, Lah R, Johnson D, Wrana A, et al. Trace element fractionation between PM10 and PM2. 5 in coal mine dust: Implications for occupational respiratory health. International Journal of Coal Geology. 2019;**203**:52-59

[36] Singh P, Tiwari D, Mishra M, Kumar D. Molecular mechanisms of heavy metal toxicity in cancer progression. In: Networking of Mutagens in Environmental Toxicology. Cham: Springer; 2019. pp. 49-79

[37] Hartwig A. Recent advances in metal carcinogenicity. Pure and Applied Chemistry. 2000;**72**(6):1007-1014

[38] Ali H, Khan E, Ilahi I. Environmental chemistry and ecotoxicology of hazardous heavy metals: Environmental persistence, toxicity, and bioaccumulation. Journal of Chemistry. 2019;**2019**:1-14

[39] Herrero M, Mendiola JA, Cifuentes A, Ibáñez E. Supercritical fluid extraction: Recent advances and applications. Journal of Chromatography A. 2010;**1217**(16):2495-2511

[40] Erarpat S, Özzeybek G, Chormey DS, Bakırdere S. Determination of lead at trace levels in mussel and sea water samples using vortex assisted dispersive liquid-liquid microextraction-slotted quartz tube-flame atomic absorption spectrometry. Chemosphere. 2017;**189**:180-185

[41] Martin AP, Turnbull RE, Rissmann CW, Rieger P. Heavy metal and metalloid concentrations in soils under pasture of southern New Zealand. Geoderma Regional. 2017;**11**:18-27

[42] Tsalbouris A, Kalogiouri NP, Samanidou VF. Recent advances in miniaturized microextraction techniques for the determination of bisphenols in environmental samples: An overview of the last two decades. Current Analytical Chemistry. 2021;**17**(4):478-494

[43] Łuczynska J, Paszczyk B, Łuczyński MJ. Fish as a bioindicator of heavy metals pollution in aquatic ecosystem of Pluszne Lake, Poland, and risk assessment for consumer's health. Ecotoxicology and Environmental Safety. 2018;**153**:60-67

[44] Pawliszyn J. Theory of solid-phase microextraction. In: Handbook of Solid Phase Microextraction. London: Elsevier; 2012. pp. 13-59

[45] Crucello J, Pierone DV, Hantao LW. Simple and cost-effective determination of polychlorinated biphenyls in insulating oils using an ionic liquid-based stationary phase and flow modulated comprehensive two-dimensional gas chromatography with electron capture detection. Journal of Chromatography A. 2020;**16**:460530

[46] Schmidt K, Podmore ID. Solid phase microextraction (SPME) method development in analysis of volatile organic compounds (VOCS) as potential biomarkers of cancer. Journal of Molecular Biomarker and Diagnosis 2015;**6**(6)

Chapter 2

Land-Based Marine Pollution: An Emerging Threat to Bangladesh

Md. Wahidul Alam

Abstract

Bangladesh is a densely populated coastal country in South Asia. The country has enormous resources in the coastal and marine environment of the northern Bay of Bengal. After the final settlement with India, Bangladesh has had an independent ocean boundary since 2014, but the country has no national policy to protect the environment from marine pollution. The chapter explores the applications of protection and prevention of the marine environment in Bangladesh from land-based marine pollution (LMP) as land-based sources have become the emerging threat to the Bay of Bengal in respect of marine pollution. Several relevant national laws and policies have been analyzed to examine the problems of LMP control in Bangladesh. This chapter also identifies the sources and effects of land-based pollutants, including the analysis of national activities, action plans, and management strategies to discover the challenges and gaps of the present regime for LMP control in Bangladesh. Finally, the chapter suggested a comprehensive approach to establishing national legislation to control LMP in Bangladesh by implementing the national and regional strategy.

Keywords: land-based pollution, marine environment, Bangladesh, legal analysis, implantation framework

1. Introduction

Bangladesh is the country of conglomerate islands in South Asia and one of the world's largest deltas. The country is mainly formed by the Ganges-Brahmaputra-Meghna (GBM) river system and consists of flat and low land. The total land area is 147,570 sq. km. It has a network of many rivers with tributaries and distributaries crisscrossing the country, whereas 405 are recognized in 'Nod-Nodi' [1]. It has a population of about 164 million, and 35.6% of the population is urban [2]. It is estimated that the country's population will be 172.4 million by 2025 [3]. Increasing trends of population increase human activity, which is dependent on the earth's environment and its natural resources system but due to industrialization and various effects of human activities cause, environmental pollution, including the marine environment. The marine environment of Bangladesh is situated at the coast of the Bay of Bengal (BOB), which is the northeastern extension of the Indian Ocean [4]. Being an extension, BOB shares many oceanic characteristics of the Indian Ocean and has active connections with the Andaman Sea, Malacca Strait, Palk Strait, etc. Bangladesh has received 118,813 sq. km. of maritime area in the BOB after the final dispute settlements with Myanmar and India in 2012 and 2014, respectively [5]. In Bangladesh, the BOB coast covers 710 km in length, and about 36 million people live in this area for their livelihoods like agriculture, fishing and aquaculture, salt farming, forest resources, and nearshore transportation. Over 90% of living and non-living resources of the world are found within a few hundred kilometers of the coast, and nearly two-thirds of the global population lives on the coast [6]. Rapid urbanization, deforestation, and unplanned extraction of marine and coastal resources aggravated the environment complex.

The ocean produces 70% of the oxygen and 80% of the world's animals and plants [7]. It is a part of the cultural and natural heritage of the world [8] and has great importance for tourism, mineral extraction, recreation and transportation, waters, beaches and cliffs, coral reefs, islands, mangrove forests, port, harbors, etc. The ocean compromises 95% of the global water supply [9], and the ocean water covers 72% of the earth's surface. Because of the importance of the world's oceans, the marine environment may have been called the 'lungs of the earth.' However, the marine environment is deteriorating due to unsustainable utilization and exploitation of ocean resources and various sources of pollutants released into the ocean. Several studies have revealed that preserving and protecting the marine environment from different deteriorating agents affects directly or indirectly like marine pollution [10]. As the marine environment is the vital diversity of marine resources, including marine animals and plants, prevention and control of marine pollution are significantly important and interlinked. Marine environmental protection is generally considered 'protection from pollution' [11], whereas pollution has a significant threat to ocean life and affects the marine species, ecosystem, and human health. Bangladesh's marine pollution is mainly caused by land-based and sea-based sources [12]. Several point and non-point sources are the leading causes of land-based sources pollution, whereas sea-based sources pollution occurs mainly due to shipping activities in the Sea [13]. Marine oil pollution can expose human health and other species to diseases [14]. Organic and inorganic pollutants from land-based, including ship-breaking activities in Bangladesh, affect the marine ecosystem essentially [15]. Although various sources of pollution affect the marine environment, several studies have indicated that the sources of land-based marine pollution (LMP) are the dominant threat in Bangladesh [16]. Hence, it has become urgent to enact legislation to regulate marine resource exploitation and utilization and marine environmental Protection regarding the sources of LMP.

Chapter 17 of Agenda 21 states about the Marine Environment Protection from LMP sources like: 'Degradation of the marine environment can result from a wide range of sources. Land-based sources contribute 70 percent of marine pollution, while maritime transport and dumping-at-sea activities contribute 10 percent each. The contaminants that pose the greatest threat to the marine environment are invariable order of importance and, depending on differing national or regional situations, sewage, nutrients, synthetic organic compounds, sediments, litter and plastics, metals, radionuclides, oil/hydrocarbons, and polycyclic aromatic hydrocarbons (PAHs). Many polluting substances originating from land-based sources are of particular concern to the marine environment since they exhibit toxicity, persistence, and bioaccumulation in the food chain. Currently, there is no global scheme to address marine pollution from land-based sources' [17]. The above section of Agenda 21 proved that prevention and control of LMP are significantly essential and linked with ocean resources conservation and management, ocean health improvement, and maritime security regarding marine environment protection. Subsequently, arguments for protecting the marine environment of Bangladesh have been put forward by devising relevant national and international legislation to achieve the goals.

2. LMP status in Bangladesh

According to National report of the Bay of Bengal Large Marine Ecosystem (BOBLME), LMP is discharged directly into the Bay of Bengal through rivers, floods, tides, groundwater, or atmospheric deposition [18]. This report also identified two kinds of land-based sources of marine pollution in Bangladesh: point and non-point (**Table 1**). A point source of pollution is those for which a particular point of entry into the environment can be identified, whereas non-point sources are harder to identify and cannot be as readily identified as originating from a single source [19]. LMP can also inflict oil sewage and industrial waste, chemical fertilizer and pesticides, water from power stations, atmospheric discharges from vehicles, chimney fumes, and sprayed agricultural chemicals [20]. In Bangladesh, industrial wastes contain high organic pollutants, and most importantly, there is no waste treatment plant inside the industries [21]. Municipal garbage is also one of the leading factors for LMP in Bangladesh due to rapid urbanization, indiscriminately dumping into the drain, canal, or river directly [21]. Agriculture activities are also the reasons for LMP as pesticides and fertilizers are carried out by rivers, canals, or drains to the sea during the rainy season [21]. The causes of land-based sources of marine pollution are the same in other countries except for ship-breaking pollution because almost all shipyard in Bangladesh is located near the coast, and their recycling operation is done through the 'beaching' method. According to the National Program of Action (NPA), to protect the coastal and marine environment from land-based activities in Bangladesh, reported 'ship-breaking pollution' is one of the critical issues for land-based marine pollution at BOB [10].

Municipal waste comprises household, industrial, commercial, agricultural, street sweeping, construction debris, sanitation residues, etc. The total amount of municipal waste generated in the country appears to be related to socioeconomic factors, and also, the total amount of solid and liquid wastes of Bangladesh that could either directly or indirectly find their way to the coast of the BOB has not been estimated yet [22]. In 2005, the average per capita municipal waste generation rate was estimated to be 0.41 kg/capita/day, and it assumed that by 2025, the total waste generation would be 47,064 tons/day with a waste generation rate (WGR) of 0.6 kg/capita/day [23].

As Bangladesh has no domestic waste treatment facilities yet, effluents from city areas directly or indirectly enter the river. Increasing human settlement in the coastal

Land-based sources of marine pollution	
Point sources	**Non-point sources**
Wastewater treatment plants Untreated sewage/outfalls Partially treated/ untreated Industrial effluent outfalls Aquaculture effluents	Agricultural runoff (pesticides, fertilizers, and animal wastes) Oil, grease, and toxic chemicals from urban runoff and energy production Coliform Bacteria and pathogens Sediments (from construction), hill cutting and deforestation. River runoff

Table 1.
Critical sources of LMP in Bangladesh [18].

areas is also the factor for increasing pollutants in the coastal environment and, in general, dumping excreta in drains and canals that go to nearby rivers and ultimately falls into the Bay of Bengal. Septic tank effluents are also dumped into the rivers and canals directly or indirectly and cause localized water pollution surrounding the drainage outfalls. The rivers, including the Karnafuly and Passur, receive raw excreta daily from dense populations on both sides of these towns. Every day a considerable amount of blood and intestines of slaughtered animals from the Firringee Bazar slaughterhouse of Chittagong find their way into the River Karnafuly and hence to the Bay of Bengal. The most common system currently adopted for the disposal of solid wastes is to dump vast quantities of the collected solid wastes on the outskirts of townships as landfills, creating a breeding ground for houseflies and mosquitoes. The decomposition process pollutes the area emitting poisonous and obnoxious smells. The major channels that carry domestic wastes and spread into the coastal city areas cause pathogenic microbial pollution and severe health hazards during the rainy season, and the severity of floods is more in the Bay [24]. Solid waste is washed by rain, the leachate mixes with the water, goes to the river, then to the Sea, and ultimately pollutes the total water body.

Pathogenic organisms are usually present in the soil and solid wastes. These pathogens are causative agents of different diseases in human beings [25]. There is a very high abundance of oligochaetes and other pollution indicators of benthic organisms near sewage outfalls in the Karnaphuli River estuary, which indicates localized pollution in the estuary [26]. High BOD indicates oxygen depletion, endangering fish and aquatic species; deoxygenation induced by toxic waste occurs in the Sea. Harmful effects of the Chittagong Urea Fertilizer Limited (CUFL) effluents on aquatic organisms, particularly on plankton and fish [27]. The pollution caused by the tannery industry, which ranks fourth in earning foreign exchange, causes phenomenal environmental pollution to the soil, ecology, and human body [28]. Especially children and aged people suffer most from the emission of sulfur dioxide gas, which causes respiratory problems, and damages the lungs of the human body [29]. Toxic metals can pass through the food chain to human beings and damage the brain and nervous system. Mercury contamination on some commercially important shrimp and finfish species from the Bay of Bengal off the coast of Bangladesh [30]. The mercury concentrations show that in estuarine shrimp, contamination is higher than in the coastal or open sea shrimps. Heavy metals like mercury are serious pollutants because they are stable compounds that are not readily removed by oxidation, precipitation, or other natural processes. Lower concentrations of methyl mercury may kill aquatic organisms.

The reduction of oxygen in water affects aquatic flora and fauna [30]. As a result of ship scrapping activities, various refuse and disposable materials are being discharged and spilled from the scrapped ships and often get mixed with soil and seawater. Due to shipbreaking pollution, there is a significant difference in the physio-chemical properties of the beach soil within and outside of the areas [31]. Phytoplankton organisms and algae suffer from reducing light intensity beneath an oil film, which inhibits photosynthesis. Coating feathers with oil, which causes buoyancy and insulation losses, sometimes cause damage to the marine bird [29]. Persistent toxic metals that settle down on the sediment from various sources threaten the survival of health of all organisms and biodiversity, besides several environmental consequences resulting from coastal aquacultures such as Shrimp farming, hatcheries operation, and shrimp and fish processing units. Encroachment and unplanned expansion of shrimp farming within the coastal belt are responsible for the destruction of forest and agricultural land, thus reducing agricultural production and endangering biodiversity. Residues in seafood, along with those in paddy-cum-shrimp culture, salt-cum shrimp culture, and several agrochemicals used in the seashore culture, ultimately go to the Sea and have an adverse impact [32].

3. Importance of LMP control in Bangladesh

Several scholars have argued that environmental quality conservation and prevention is the cry of the day [33]. During the last few centuries, with the growth of the population, the environment has been subject to harsh treatment by various human activities [33]. Around 1.8 metric tons (MT) of pesticides enter the BOB, and Bangladesh imports approximately 3.5 million tons of refined and crude oil from different countries, contributing to around 4–6 MT of oil pollution in the BOB [34]. According to World Casualty Statistics 2011, the largest five-ship scrapping countries in the world are India, Bangladesh, Pakistan, China, and Turkey, which recycle 97–98% of the world's tonnage [35]. Presently, Bangladesh is the first largest country for ship scrapping worldwide [36]. Each ship released more than 250 kg of polychlorinated biphenyl (PCB) in the BOB, whereas 194 ships were dismantled in 2015 on the Bangladesh coast [37]. This toxic, hazardous waste threatens marine and coastal environments and public health. Marine microbial pollution also occurs due to the excess discharge of city garbage and industrial waste, which helps grow health hazard pathogenic bacteria in the marine environment [38, 39]. Bangladesh has no domestic waste treatment facilities; therefore, all townships and human effluents, directly and indirectly, fall into the river with the untreated condition and hence to the BOB through the estuary. All the coastal city areas carry household wastes and cause pathogenic microbial pollution and severe health hazard diseases for coastal residents [26]. Although a variety of sources of pollution affect the marine environment and the sources of LMP are the dominant threat in Bangladesh [34] so, it has become an urgent task to enact legislation to regulate marine resource exploitation and utilization as well as marine environmental protection regarding the control of LMP. Bangladesh has ratified most international conventions regarding marine pollution to protect the marine environment but has yet to initiate domestic laws to affect international commitments. Bangladesh has no domestic law regarding marine pollution control or prevention except 'Territorial Waters and Maritime Zones Act (TWMZA'). In the context of maintaining the ecological balance in the Bay of Bengal (BOB), the TWMZA) adopted in 1974, the only act regarding the marine sector in Bangladesh, but there were no provisions to control marine pollution. Section 8 of this Act implies that 'the government may take the initiative to make comprehensive rules regarding marine environmental pollution' [40]. As Bangladesh's final settlement of maritime boundary delimitation issues finished in 2014 with the ITLOS verdict, it is time to comply with the duty to protect the marine environment and adopt laws and regulations to control, reduce and prevent the sources of LMP in Bangladesh.

Globally, Marine environmental protection regarding the LMP threat is also an emerging issue as marine pollution has devastating impacts on the marine environment. Efficient management, comprehensive regulation, and regional cooperation are necessary to overcome the LMP threat [41]. Before 1954s, there was no convention to combat marine pollution; no emphasis was placed on the importance of LMP control [41]. At last, in the 1960s, the Paris Convention relating to LMP control [42] was negotiated, and other multilateral conventions, including the Helsinki Convention, were adopted with the provisions of LMP control [41]. The [43] was the most important, updated, and comprehensive international agreement regarding previous marine environmental protection laws [44, 45], which provides a comprehensive framework for protecting and preserving the marine environment [43]. UNCLOS contains various sources of marine pollution, including LMP, and provides a framework for developing and conserving marine living resources [44, 46]. According to Chapter 17, Agenda 21, UNCLOS is a necessary tool for developing and utilizing international and regional agencies regarding ocean pollution prevention [47]. The United Nations Conference on Environment and Development'1992 (UNCED) also considered the LOSC a significant contribution to marine Protection and International Law. The United Nations (UN) Secretary-General states about the marine environmental Protection regarding international Law at the conference of UNCED'1992; The UNCLOS 'provides a model for the evaluation of International Law and its incorporation of several newly developed concepts and principles, such as the prevention of Transboundary pollution; assessment; habitat protection; and eco-system considerations; an integrated approach to the various source of pollution; and contingency planning against pollution emergencies' [48]. As Article 207 and 213 of the UNCLOS provides, state shall prescribe and enforce legislation to reduce, prevent and control LMP [49], which is the only global treaty with special provisions for LMP. In advance, Montreal Guidelines were adopted with a comprehensive management framework to protect LMP, and the Global Program of Action (GPA), 1995 represents the significant development regarding global arrangements of LMP control [50].

According to Sir Edmund Hillary, 'Environmental problems are social problems. They begin with people as the cause and end with people as victims' [33]. As I mentioned, land-based pollution has been subject to harsh treatment by various human activities, and we have made this by our irrational and selfish conduct, so all these environmental problems and crises have to be analyzed and judged in the context of LMP control. Hence, the development and implementation of legislation are required to protect the marine environment regarding LMP control in Bangladesh. Ensuring good governance in the Sea, promoting economic growth, ensuring the security of maritime recourses from pollutants, improving the value of the maritime environment and ensuring their sustainability, adoption of a comprehensive Land-Based Marine Pollution Control Act (LMPCA) for Bangladesh is urgently needed with the implementation analysis of regional and international cooperation. LMPCA can be a safeguard for the sustainable Protection of the marine environment in Bangladesh.

4. Present status of responsible national organizations in Bangladesh

Enacting enabling legislation nationally is vital, but implementing domestic legislation remains the major challenge. Indeed, several institutions are responsible for protecting the marine environment of Bangladesh. The Ministry of Shipping (MoS) is the lead ministry, and the Department of Shipping (DoS) is the lead agency dealing with vessel-source pollution. This department is aided by several subordinate offices under its administrative control, including the mercantile marine department, marine academy, government shipping office, and seamen's training school [51]. The ports of Bangladesh, Chittagong, and Mongla were established by a different law and treated as autonomous bodies. Other institutions like the coast guard, DoE, and Bangladesh Navy (BN) are entrusted with respective responsibilities to protect the marine environment. Besides the association of social development, the Bangladesh

Name	Roles
Ministry of Shipping Department of Shipping Mercantile Marine Department Government Shipping Office Marine Academy Seamen's Training School	Implementation of all IMO Conventions Liaison with IMO and other International Organizations Inspections of ships Enforcement of Merchant Shipping Ordinance, 1893, and all other relevant legislation. Registration, survey, and certification of ships. Training and certification of marine officers, engineers, and seamen.
Chittagong and Mongla Port Authorities	Management and development of ports Prevention of pollution in the port area.
Coast Guard	Patrolling in the maritime area of Bangladesh. Detection of activities causing pollution of the environment in the maritime zones of Bangladesh and taking measures for their stoppage Enforcement of any warrant or any other order of any court or other authority in respect of any ship, which has entered the territorial waters of Bangladesh. (Section 7 of the Coast Guard Act 1994)
Bangladesh Navy	Safeguarding the country's sovereignty over the internal waters and territorial sea, and sovereign rights over the Contiguous Zone, Exclusive Economic Zone (EEZ), and Continental Shelf.
Department of Environment	Overall management and protection of the environment Enforcement of Environment Conservation Act 1995

Table 2.
Key government agencies involved in the prevention of marine pollution.

Environmental Lawyers Association (BELA), the Center for Environmental Studies and Disaster Management, the Centre for Environmental Studies and Research, and Dhaka, Chittagong, Khulna Universities, and Bangladesh Agriculture universities are also engaged in marine environmental research, monitoring and enforcement opportunities (**Table 2**) [52].

The Ministry of Environment, Forest and Climate Change (MOEFCC) ensures a sustainable environment and optimum forest coverage through ecosystem and biodiversity conservation; controlling environmental pollution; addressing climate change; research, floristic survey, and developing forest resources. Section 3 of the Bangladesh Environment Conservation Act, 1995, establishes the DoE. The headquarters of the environment department is currently organized into two main functional areas 'Administration, Planning and Development, and 'Technical' [53]. A director heads each, and there is a lack of knowledge or professionalism regarding marine environment protection. They have no data or do not provide proper knowledge regarding the current status of marine pollution and their mitigation steps.

The MoS is the highest governmental institution in Bangladesh to implement IMO Conventions in Bangladesh. The ministry encompasses within its fold shipping and port sectors, including national waterways, inland water transport, ports, and ocean shipping, and oversees the safety and environmental matters and the regulatory aspects of maritime shipping and maritime education. The ministry is responsible for formulating policies and plans for these subjects and facilitating the quick implementation of various projects. Besides, this ministry also cares about maintaining and expanding viable, efficient, and dependable water transportation and communication systems as the cheapest way of economic activities in rural and urban areas. The DoS is an agency under the MoS, Bangladesh. Bangladesh's maritime safety administration is responsible for formulating and implementing the national policies and

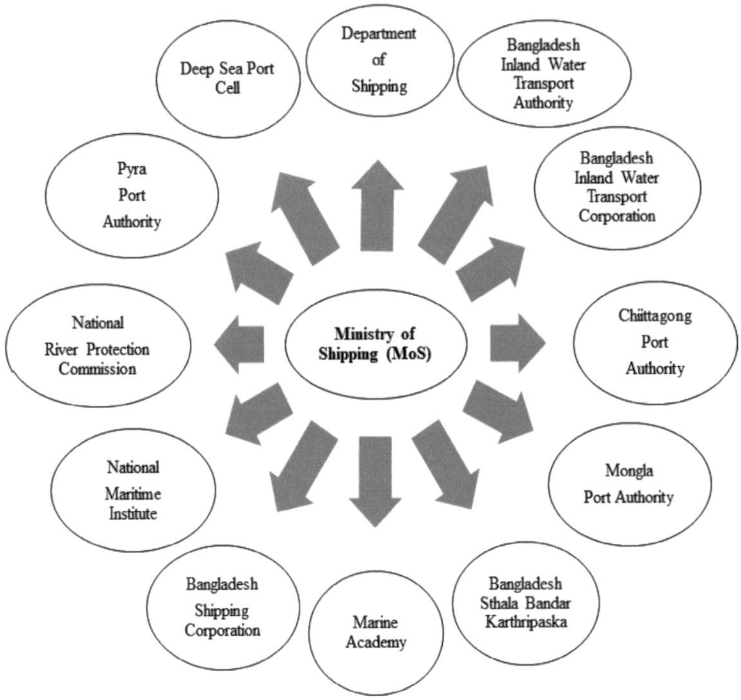

Figure 1.
Organization under the Ministry of Shipping.

legislation to ensure the safety of life and ships at Sea, development of the shipping industry, maritime education and certification, employment and welfare of seafarers, and other shipping-related matters. The department is also responsible for ensuring compliance with international conventions relating to maritime matters (**Figure 1**).

Bangladesh Oceanographic Research Institute (BORI) has started its journey at Cox'sbazar under the Ministry of Science and Technology (MoST) for all types of oceanographic research, including environmental impact assessment and oil-spill risk assessment. A law named 'Oceanographic Research Institute of Bangladesh 2015' (Act No. 7 of 2015) has been passed in the parliament. This act provides for the creation of the Bangladesh Oceanographic Research Institute. It establishes the composition, duties, and responsibilities of the Institute mentioned above, entitled to undertake the research and disseminate related information results and activities of all concerned operations, management, and control of the Bangladesh oceanographic sector. Chittagong and Mongla seaport is considered the center of the economy of Bangladesh. Both port authorities were established under the Chittagong and Mongla Port Ordinance 1976, amended in 1995. Recently, the Chittagong port authority initiated Environment Management Unit (EMU), which will focus on protecting the marine environment, but it is yet to get approval from the ministry. The BN is entrusted with the responsibility to protect the maritime boundary of Bangladesh. The BN is mainly limited to coastal patrolling, yet it is an enormous scope to protect the marine environment from pollution. Bangladesh Coast Guard is the principal Law enforcement authority that implements national and international maritime laws. However, the Coast Guard of Bangladesh deals with only piracy operations rather than marine environmental Protection. Mercantile Marine Department was established under the Bangladesh Merchant Shipping Ordinance 1983. The department has an essential responsibility to deal with the accidents that may cause marine pollution, but it lacks effectiveness in dealing with the issue.

5. An emerging threat and regional activities on LMP control

Generally, a regional approach is based on the mechanisms of interest, traditions, and values of neighboring countries [54] under common interests or problems. It also depends on political, social, economic, and cultural cooperation and interaction [54]. Marine environment pollution in the Bay of Bengal sub-region is a common problem because every year, this Bay receives many pollutants from land-based and other activities [55]. These pollutants reach the marine and coastal waters through rivers, canal systems, or the atmosphere [55]. For example, it is estimated that in Delhi, 5 million gallons of water containing DDT wastes are dumped into the Jamuna river daily, which ultimately falls through the Bay of Bengal. The South Asian Seas Action Plan, Colombo Workshop on LMP Control, and Bay of Bengal Large Marine Ecosystem (BOBLME) program are vital regional approaches for researching marine environmental Protection from pollution directly or indirectly in this region.

South Asian Seas Action Plan (SASAP) South Asia is one of the most diverse regions in the world. They are bordered to the north by the Himalayas and to the south by the Indian Ocean. Many countries have taken action to protect and manage the environment. They are also parties to many multilateral environmental agreements requiring them to work cooperatively to mitigate concern issues. The South Asian Seas Action Plan (SASAP) is a Regional Action Plan (RSP) under the UNEP Regional Seas Program (RSP) adopted in 1995. The Regional Seas Programs of UNEP have several common elements. Establishing a Regional Seas Program usually begins with developing an action plan outlining the strategy and substance of a regionally coordinated program to protect a typical body of water. The action plan is based on the region's environmental challenges and socio-economic and political situation. It may cover issues ranging from chemical wastes and coastal development to marine species and ecosystem conservation.

In most cases, the action plan is underpinned by a solid legal framework in a regional convention and associated protocols on specific problems. The legally binding convention expresses governments' commitment and political will to tackle common environmental problems through joint, coordinated activities. The South Asia Cooperative Environment Program (SACEP) has been privileged to participate in developing the RSP in the SASAP.

The overall objective of the SASAP is to protect and manage the region's marine environment and related coastal ecosystems in an environmentally sound and sustainable manner. In addition to specifying the needs under the main components of Environmental Assessment, Environmental Management, Environmental Legislation & Institutional, and Financial Arrangements, the Action Plan identified the areas where priority activities need to develop for implementation. The action plan is designed to develop financial and institutional mechanisms for protecting marine and coastal environments from different activities, including LMP control. The Plan focuses on Integrated Coastal Zone Management (ICZM), oil-spill contingency planning, human resource development, and land-based activities' environmental effects. For LMP control, they suggest a regional action plan for reducing nutrient loading into the coastal waters of the South Asian Seas Region. The key activities of the action plan included Inventory of point or non-point sources of nutrients that end up in the coastal waters, developing and undertaking actions to reduce nutrient inputs to agriculture, and developing a regional action plan to pursue by member countries [56]. Although there is no regional convention, SASAP follows existing global environmental and maritime conventions and considers the Law of the Sea's umbrella convention.

5.1 Colombo workshop on LMP control

On 22–25 October 1997, the UNEP organized a workshop on the Implementation of the GPA for the Protection of the Marine Environment from LMP in the South Asian Regions, also known as 'Colombo Workshop on LMP Control.' The workshop deliberated a draft about the overview of land-based sources and activities that affect marine, coastal and freshwater environment, regional action program, and their implementation process, including institutional activities, financial consideration for the Protection of the marine environment from land-based activities in the South Asian Seas (SAS) [57, 58]. As India, Maldives, Nepal, Srilanka, and Pakistan are the representative countries of the Workshop and SACEP acted as Secretariat for the Workshop, the participants need special attention to protect the marine environment from LMP control. The Colombo workshop also suggested an integrated approach to developing national action programs to implement national environmental strategies and policies, including a regional action plan and financial cooperation among the related stakeholders [57]. However, Bangladesh was not attending this workshop so that Bangladesh could follow the recommendations of the Colombo workshop for effective involvement with GPA in the control of LMP in the SAS region.

5.2 Bay of Bengal Large Marine Ecosystem (BOBLME)

The BOBLME Project has been working closely with various bodies and organizations operating in the Bay of Bengal to improve the coordination of activities and enhance impacts. Bangladesh, India, Indonesia, Malaysia, Maldives, Myanmar, Sri Lanka, and Thailand are collaborating through the Bay of Bengal Large Marine Ecosystem (BOBLME) Project to better the lives of their coastal populations by improving regional management of the Bay of Bengal environment and its fisheries. A proposal was made to develop a strategic action program (CAP) to protect the ecosystem's health and manage the living resources of the Bay of Bengal. SAP-based on the Transboundary Diagnostic Analysis (TDA) was endorsed in March 2012 by the BOBLME countries. The TDA identifies the main transboundary issues and their causes. It reviews the driving forces at work in the BOBLME, such as the socioeconomic, institutional, legal, and administrative circumstances and the projected impact of climate change on the region. The negotiated policy document sets out a program of actions that address the causes of the major fisheries, environmental, social, and economic issues. However, the program did not directly relate to LMP control. However, they aimed to assist the countries of this sub-region in meeting their obligations to manage the marine and coastal environment, which is directly related to SASAP, GPA, etc. Therefore, this program is indirectly relevant to LMP control in this region [41].

6. Challenges in dealing with such an emerging threat in Bangladesh

Degradation of the marine and coastal environment can result from various sources. A precautionary and anticipatory rather than reactive approach is necessary to prevent the further degradation of Bangladesh's coastal and marine environments. These require the integration of social and economic components along with environmental components. Environmental concerns, experimental programs, sustainable livelihood, economic considerations, poverty reduction, coastal development, and other stakeholders' involvement are the key objectives of the national management strategy for LMP control in Bangladesh [10].

Innovative strategies and practices compatible with environmental sustainability are required to protect Bangladesh's overall coastal and marine environment from land-based activities. Developing and implementing conservation practices aiming at harmonizing the efforts from different sectors with environmental conservation as the ultimate goal targeted. Strengthening and upscaling the ongoing conservation program activities and strategies are necessary. Moreover, efforts should have been made to protect the marine ecosystems of Bangladesh from further degradation by proposing and improving the application of legal instruments to protect the marine environment and strengthen existing control systems. Large-scale shrimp farming and disposal contribute to high coastal and marine pollution. Organic shrimp farming should practice and develop a sustainable model like the mangroves—aquaculture. The United Nations Environment Program (UNEP) formulated its 1995 GPA to protect land-based activities' coastal and marine environment. In 2003, a meeting was held in Sri Lanka on GPA for South Asia Region. Several projects like coastal mangrove ecosystem, adverse impact on the marine environment and human health, shipbreaking activities, Coastal fisheries management, solid waste management, environmental flow requirement, assessment, etc., are funded by UNEP/GPA at an experimental scale.

Marine and coastal protection strategies would necessitate interventions to improve people's capacity for sustainable livelihood. To conserve and preserve natural resources, including agriculture and fisheries, management activities would include education, training, and awareness of the local people to sustainable production concepts. To improve coastal water quality, the Poverty Reduction Strategic Paper (PRSP) set the strategic goals by installing (Effluent Treatment Plant) ETP in all industries, including shipbreaking. To ensure the proper utilization of resources and comprehensive development of the coastal and marine areas, the Bangladesh government approved the 2005 'Coastal Zone Policy'. The policy focused on the economic development of the coastal area, including livelihood improvement, mitigation, coastal resources management, land utilization planning, Protection of salinity intrusion, etc.

The MoEFCC of Bangladesh has developed relevant environmental laws like Environment Policy 1992, Environment Conservation Act 1995, Environment Conservation Rules 1997, and Environment Court Act 2000. It has also set up offices at the divisional level to implement environmental rules and regulations. It appears that the existing institutional capacity and allocated resources are inadequate compared to the scale and magnitude of the prevailing environmental problem. Addressing environmental problems related to land-based activities in Bangladesh is new and institutional collaboration and integration with other sectoral programs are limited.

Globally, LMP control has specific problems due to geographical location, ecological factors, social structure, economic activity, and scientific evidence [59].

Besides the lack of scientific and technical data, industrial production restrictions on sovereignty are also fundamental obstacles to controlling LMP [41]. Besides the concept of marine pollution and its impacts, lack of legislation, separate marine environmental protection department, political wellness, an unstable political situation, lack of coordination of the government officials, institutional incapacity, etc., make the challenges to control LMP in Bangladesh. This section also discusses the scientific, economic, and transboundary problems and issues to control the LMP in Bangladesh. Data collection, source identification, and damage determination from land-based sources are the main elements of scientific problems. Industrial production versus marine environmental pollution issues are elements of economic problems. Types of contaminants, regional consensus, and social priorities are the main factors for transboundary problems.

7. Suggested framework

In 2004, the Department of Shipping (DOS), Bangladesh drafted a 'Marine Environment Conservation Act 2004,' which has not been ratified yet. It hoped that the government would enact comprehensive maritime environmental protection laws and regulations, including LMP control, to protect the BOB from land-based sources of pollution. In 2006, the National Program of Action (NPA) under the United Nations Environment Program (UNEP) identified 12 major issues that are the critical sources of land-based pollution at BOB. Generally, climate change, industrial waste (including ship breaking yards), land-use change, deforestation, coastal zone erosion, sewage disposal, agrochemicals, solid waste management, salinity intrusion, rapid urbanization, and coastal tourism is the main creature of pollution in the marine environment. To prevent LMP, international soft laws, complex laws, and conventions are available to deal with marine environmental pollution worldwide. However, Bangladesh ratified some of the laws and conventions but has not yet enacted any law regarding marine pollution control due to a lack of proper national legal framework and willingness of the government; these conventions have not yet been ratified. On the other hand, Bangladesh cannot fully implement international marine environmental pollution related to legal issues due to a shortage of financial reasons [21]. As Bangladesh has an ocean boundary but no marine policy to restore the ocean resources at BOB, it is necessary to set up a marine policy emphasizing LMPCA.

LMPCA framework for Bangladesh can be developed based on active approach, responsibility, indigenization, standards, norms, practices, monitoring, assessment, legal compliance, etc. The layout and contents of the LMPCA can be categorized into four segments. The first segment covers the mission, vision, objectives, goals, and guiding principles. The second segment should contain the LMPCA domain discussing sea area, geography, resources, economy, sources, and impacts of marine pollution with an example. The third segment contains the proposed framework's challenges, implementation, and monitoring mechanisms. The last segment sets out detailed policy guidelines for all stakeholders and leads actors. The Ministry of Environment and Forest (MOEF) and the Ministry of Shipping (MOS) can jointly create legislation for favorable economic and environmental conditions in ship recycling areas by implementing hazardous waste treatment facilities and ensuring logistic support for enforcing Law and order against marine pollution. Besides, the Department of Tourism ensures the sea beaches management and formulation of strategies to strengthen eco-tourism. Department of Fisheries (DOF) carries out R&D in marine fishing. Department of Science and Technology (DOST) undertakes research and projects to conserve marine landscapes and ecosystems for sustainable management of marine living resources from marine pollution.

The MoEFCC, MoS, and other related departments (**Figure 2**) should jointly advise the act's functions. The policy guidelines will be standard for their sectors but preliminary. The policy can follow the TWMZ act, 1974, and MEC act, 2004, to organize the policy structure. Comprehensive policy guidelines for marine environment protection from marine pollution are set out for all sectors and stakeholders related to maritime activities. So all the guidelines are directed towards the lead ministries where some are very common for all, and some are specific to marine environment protection. MoEF takes all measures to prevent pollution and Implements laws and

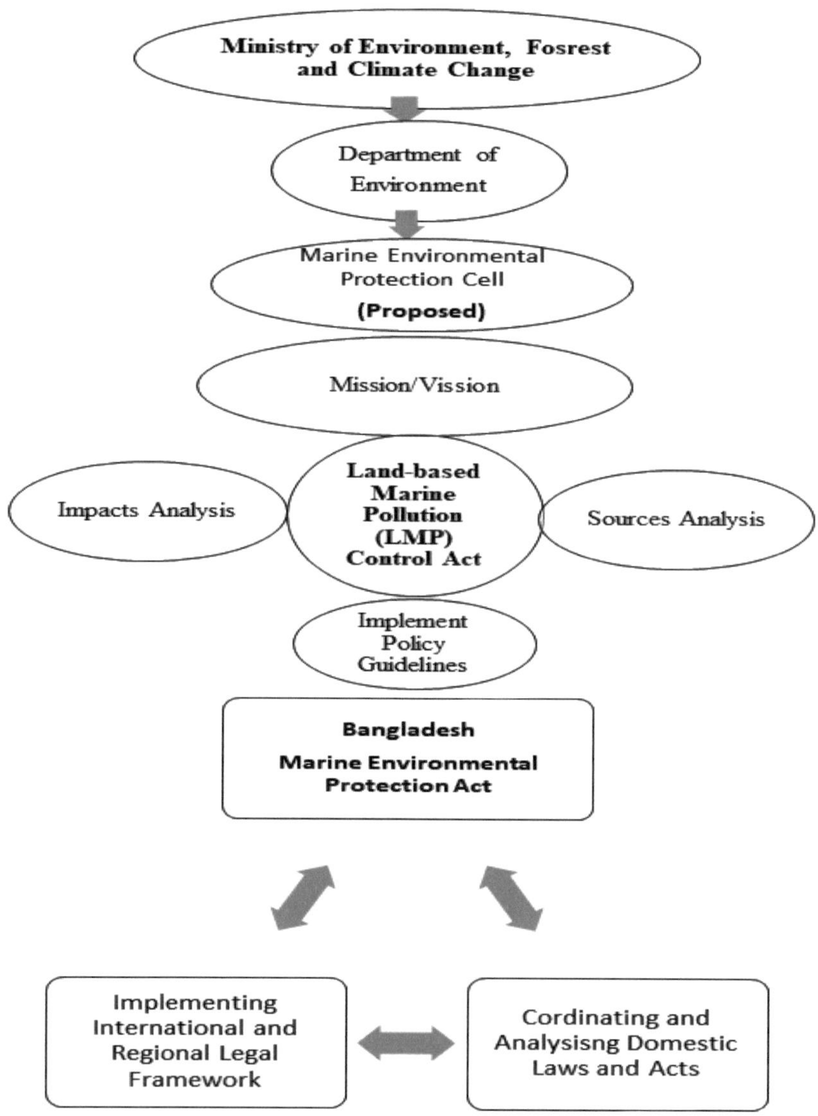

Figure 2.
A suggested framework for implementing LMPCA in Bangladesh.

regulations to protect the marine environment from land-based sources and sea-based activities. This ministry also develops a marine environmental and coastal protection strategy and conservation. MoS take measures to develop an eco-transport network during the seaborne trade. Create a conductive eco-friendly policy for flourishing private shipping industries regarding marine environmental pollution. The ministry also takes adequate measures to ensure the environmental safety of Bangladeshi vessels.

8. Conclusion

Bangladesh ratified the [43] on 27 July 2001. Bangladesh needs to enact a comprehensive law on marine pollution because there is no uniformity among the national laws; moreover, penalties are different for the same offenses in different laws. Bangladesh has plenty of mineral resources in the coastal marine area. However, there is no law regulating Bangladesh that defines and incorporates provisions on marine pollution, and the government should initiate effective mechanisms to solve this serious problem. To implement international legislation on land-based sources of marine pollution, the country should immediately undertake capacity-building projects for its marine and port administration. A budget section should be allocated on a priority basis to establish reception facilities and collect pollution detection equipment.

Moreover, the government may wish to seek private partnerships to provide reception facilities. The government should enact a comprehensive law focused on protecting the marine environment, including LMP control or, more specifically, the Land-based Marine Pollution Control Act (LMPCA), as land-based pollution is dominant in Bangladesh. However, Bangladesh has enormous resources in the coastal and marine areas, so the Protection of this region should be increased through the proper legislative framework. Furthermore, there are no rules to govern solid waste and land-dumping, which are the critical causes of LMP. Improving international cooperation, regional frameworks, public participation, and awareness would have to be built up to a high pitch of sensitivity for controlling land-based marine pollution to ensure a sustainable environment.

Author details

Md. Wahidul Alam
Department of Oceanography, University of Chittagong, Chattogram, Bangladesh

*Address all correspondence to: wahidul.alam@cu.ac.bd

References

[1] BWDB. Bangladesh Nod-Nodi (Rivers of Bangladesh). Dhaka: Bangladesh Water Development Board (BWDB); 2012

[2] World Meters. Bangladesh Population. Bangladesh: World Meters (WM); 2007

[3] World Bank. Bangladesh Climate Change and Sustainable Development. Dhaka: World Bank (WB); 2000

[4] IHO. Limits of Oceans and Seas. 3rd ed. Mote Carlo: International Hydrographic Organization (IHO); 1953

[5] MOFA. Press Statement of the Honble Foreign Minister on the Verdict of the Arbitral Tribunal/PCA. Dhaka: Ministry of Foreign Affairs (MOFA); 2014

[6] Alam MK. Bangladesh's Maritime Challenges in the 21st Century. Dhaka: Pathak Shamabesh; 2004

[7] Larson DL. Foreword. Ocean Development and International Law. 1982;**11**(1-2):2

[8] Kelleher G, Kenchington R. Guidelines for Establishing Marine Protected Areas; 1991. p. 8. Available at online: https://www.sprep.org/attachments/pipap/guidelines_for_establishing_MPAs_kelleher.pdf

[9] Merle PA. Deep-Sea Mining and the Law of the Sea. Martinus: Nijhoff Publishers; 1983. p. 3

[10] DoE. Bangladesh: National Program of Action for Protection of the Coastal and Marine Environment from Land-Based Activities. Dhaka: Department of Environment (DoE); 2006

[11] Rose G. Protection and Conservation of the Marine Environment. Wollongong: University of Wollongong; 1996. p. 151

[12] Derraik JGB. The pollution of the marine environment by plastic debris: A review. Marine Pollution Bulletin. 2002;**44**(2002):842-852

[13] UNEP. An Overview of Land Based Sources of Marine Pollution. Jamaica: United Nations Environmental Program (UNEP)-Caribbean Environment Program; 2001

[14] EPC. Oil Spills' Effects on Human Life. Environmental Pollution Center (EPC); 2018. Available at online: https://www.environmentalpollutioncenters.org/oil-spill/humans/

[15] Hossain MM, Islam MM. Ship Breaking Activities and Its Impact on the Coastal Zone of Chittagong, Bangladesh: Towards Sustainable Management. Chittagong: Young Power in Social Action (YPSA); 2006

[16] Francis J, Mmochi AJ. Land Based Activities and Sources of Pollution to the Marine, Coastal and Associated Fresh Water Ecosystems in the Western Indian Ocean Region. Zanzibar: Institute of Marine Science (IMS); 2003

[17] UNCED. Chapter 17 of section 17.18 at Agenda 21. United Nations Conference on Environment and Development (UNCED). It is also known as the Earth Summit. 1992

[18] Hoq ME, Haroon AKY, Hussain MG. Ecosystem Health and Management of Pollution in the Bay of Bengal. Support to Sustainable Management of the Bay of Bengal Large Marine Ecosystem (BOBLME) Project, Bangladesh Fisheries Research

Institute, Bangladesh. 2011. p. 84. Available at online: http://www.boblme-bangladesh.org/publications/reports/Ecosystem%20health%20&%20management%20of%20pollution%20in%20the%20BoB.pdf

[19] GESAMP. Protecting the Oceans from Land-Based Activities: Land-Based Sources and Activities Affecting the Quality and Uses of the Marine, Coastal and Associated Freshwater Environment. Joint Group of Experts on the Scientific Aspects of Marine Environment Protection; 2001. Available at Online: http://www.gesamp.org/site/assets/files/1251/protecting-the-oceans-from-land-based-activities-en.pdf

[20] Brubaker D. Marine Pollution and International Law: Principles and Practice. London, and Florida: Belhaven Press; 1993

[21] Ahmed S. Preventing Marine Pollution in Bangladesh: A Critical Appraisal of a National and International Legal Framework. Chittagong: University of Chittagong; 2013. pp. 1-92

[22] Kaly UL. Review of Land-Based Sources of Pollution to the Coastal and Marine Environments in the BOBLME Region. FAO-BOBLME Program; 2004. Available at Online: https://www.boblme.org/documentRepository/Theme_%20Land%20Based%20Pollution%20-%20%20Urusla%20Kaly.pdf

[23] Iftekhar E, Hashmi QSI. Community-Based Solid Waste Management through Public Private-Community Partnerships: Experiences of Waste Concern in Bangladesh. In: Third Asia Conference. Tokyo; 2006

[24] Bhuyan S, Islam S. Status and impacts of industrial pollution on the Karnafully River in Bangladesh: A review. International Journal of Marine Science. 2017;7(16):141-160

[25] Alam MW. Health Hazard Bacteria in Water and Sediment: Karnafuly River Estuary. Saarbrucken, Germany: LAP LAMBERT Academic Publishing; 2012. pp. 1-95

[26] Hossain MM. Macro-zoo benthos as pollution indicators. A quarterly Newsletter of Environmental Sanitation Information Center, AIT, Bangkok. 1987;**9**(1):8-10

[27] Hossain TM. Study on the Environment Impact Assessment (EIA) of the Effluent Discharged by the Chittagong Urea Fertilizer Limited (CUFL) on the Karnafuli River Estuary. Chittagong: Institute of Marine Science, University of Chittagong; 1992. p. 56

[28] Hossain MM. National Report of Bangladesh on sustainable Management of the bay of Bengal large marine ecosystem (BOBLME). 2004

[29] Gain P, Moral S, Raj P, Sircar L. Bangladesh Environment: Facing the 21st Century. Dhaka: Society for Environment and Human Development (SHED). 2002. p. 340

[30] Hossain MM. The concentration of total mercury in marine shrimp and fishes from Bangladesh. Journal of Marine Biology. 1994;**3**:27-32

[31] Islam KR, Hossain MM. Effect of ship scrapping activities on the soil and sea environment in the coastal area of Chittagong, Bangladesh. Marine Pollution Bulletin. 1986;**17**(3):462-463. DOI: 10.1016/0025-326X(86)90836-2

[32] BAPA. National Consultation 2002 on Bangladesh Environment & Agriculture. Bangladesh: Sustainable

Agricultural Working Group of Bangladesh ParibeshAndolon (BAPA); 2002

[33] Chakraborti NK. Environmental Protection and the Law. New Delhi.: Ashish Publishing House; 1994. p. 13

[34] Alam MW, Qayum S, Hasan MM, Xiangmin X. Land-based marine pollution control in Bangladesh: A suggested framework with a critical analysis of national legal issues. Indian Journal of Geo-Marine Sciences. 2018a;**47**(10):1909-1917. Available at online: http://nopr.niscpr.res.in/bitstream/123456789/45171/1/IJMS%2047(10)%201909-1917.pdf

[35] Daily Star. Maritime environmental laws are needed to combat marine pollution. The Daily Star. 2014. http://recyclingships.blogspot.com/2014/10/maritime-environmental-laws-needed-to.html

[36] Guardian. This is the world's cheapest place to scrap ships – But in Chittagong, it's people who pay the price. The Guardian. 2017. https://www.theguardian.com/global-development/2017/dec/02/chittagong-shipbreaking-yards-legal-fight

[37] Sarraf M, Stuer-Lauridsen F, Dyoulgerov M, Bloch R, Wingfield S, Watkinson R. Ship Breaking and Recycling Industry in Bangladesh and Pakistan. Washington, DC: The World Bank; 2010. Available at online: https://fid4sa-repository.ub.uni-heidelberg.de/3749/

[38] Alam MW, Xiangmin X, Yingjie M, Qayum S. Suggested legal framework for preventing shipbreaking pollution at Chittagong coast, Bangladesh, with analysis of relevant international issues. Indian Journal of Geo-Marine Sciences. 2018b;**47**(4):752-758

[39] Alam MW, Rahman MM. Comparative study of Salmonella and Vibrio spp. distribution in the Karnafuly River estuary in Bangladesh in response to microbial pollution. Indian Journal of Geo-Marine Sciences. 2018;**47**(05):1075-1080

[40] TWMZA. Section 8, Territory Water and Maritime Zone Act. 1977

[41] Hassan SMD. Protecting the Marine Environment from Land-Based Sources of Marine Pollution: Toward an Effective Cooperative International Arrangement. Faculty of Law, University of Wollongong; 2002. pp. 1-393. Available at online: https://ro.uow.edu.au/cgi/viewcontent.cgi?filename=0&article=2859&context=theses&type=additional

[42] ILM. The Paris Convention for the Prevention of the Marine Environment from Land-Based Sources. 1974

[43] UNCLOS. Article 207/222. United Nations Convention of Law of the Sea. 1982

[44] Card JC. Implementing the United Nations convention on the law of the sea: A coastguard perspective. The Georgetown International Environmental Law Review. 1995;7:726

[45] Kindt JW. Marine Pollution and the Law of the Sea. New York: Williams Hein and Co. Inc; 1986. p. 708. Available at online: https://www.wshein.com/media/brochures/9176.pdf?d=20221227

[46] Chenry JI. The Protection of Marine Environment by the 1982 United Nations Conventions on the Law of the Sea, Quoted in Card J C., 1995. 'Implementing the United Nations Convention on the Law of the Sea: A coastguard Perspective. The Georgetown International Environmental Law Review. 1995;7:734-735

[47] Juda. IMCO and the regulation of ocean pollution from ships. International and Comparative Law Quarterly. 1977;**26**:558

[48] Cherry JI. The marine environment and 1982 United Nations convention on the law of the sea. International Lawyer. 1994;**28**(4):888

[49] Churchill RR, Lowe AV. The Law of the Sea. 3rd ed. New York: Juris Publishing; UK: Machester University Press; 1999. p. 379

[50] UNEP. Global Program of Action for the Protection of the Marine Environment from Land-Based Activities. Washington, DC: United Nations Environmental Program; 1995

[51] Karim S. Implementation of the MARPOL convention in Bangladesh. Macquarie Journal of International and Comparative Environmental Law. 2009;**6**:51-82

[52] Holmgren DS. An Environmental Assessment of the Bay of Bengal. Madras: Bay of Bengal Program; 1994. p. 121

[53] Hasan MI. International Environmental Law: Bangladesh Perspective. Dhaka: AinPrakashan; 2014. p. 503

[54] Bennet AL. International Organizations: Principles and Issues. New York: Prentice Hall Inc; 1977. p. 289

[55] ADB. Asian Development Outlook. Manila: Asian Development Bank (ADB); 1991. pp. 221-222. Available at online: https://www.adb.org/sites/default/files/publication/180564/ado-1991.pdf

[56] BOBLME. Nutrient Loading and Eutrophication of the South Asian Seas Coastal Waters – A Scoping Study. India: BOBLME; 2015

[57] UNEP. Report of the Workshopworkshop on Implementation of the Global Program of Action for the Protection of the Marine Environment from Land-Based Activities in the South Asian Regions. Colombo: United Nations Environmental Program; 1977

[58] UNEP. Twenty-third session of the governing Council of the United Nations Environment Program: Item 4 of the provisional agenda, United Nations. 2005

[59] Dahl AL. Land-based pollution and integrated coastal management. Marine Policy. 1993;**17**(6):561-572

Section 2

Marine Pollution with Oil Spill and Advanced Monitoring and Remediation Strategies

Chapter 3

Recent Advances in Oil-Spill Monitoring Using Drone-Based Radar Remote Sensing

Bilal Hammoud and Norbert Wehn

Abstract

Oil spills are regrettably common and have socioeconomic implications on communities and disastrous consequences on the marine ecosystem and maritime life. The European Space Agency (ESA) has stated that worldwide spillage exceeds 4.5 million tons of oil annually, where 45% of the amount is due to operative discharges from ships. To alleviate the severity of oil spills and promptly react to such incidents, it is crucial to have oil-spill monitoring systems, which enable an effective contingency plan to dictate the best actions for dealing with oil spills. A quick and efficient intervention requires the (1) detection of oil slicks, (2) thickness estimation, and (3) oil classification. The European Maritime Safety Agency (EMSA) highlighted in 2016 the need to use drones as complementary systems supporting satellite maritime surveillance. While multiple sensors could be used, active radars appear to be prominent for oil spill monitoring. In this chapter, we present recent advances in drone-based radar remote sensing as an effective oil spill monitoring system. It shows from the system-level perspective the capability of radar systems on drones, using high spectral resolution and parallel scanning, to perform the above-required functionalities (1, 2, and 3) and provide valuable information to contain the damage.

Keywords: oil spills, drone, radar, reflectivity, slick detection, thickness estimation, classification

1. Introduction to oil spills: Sources, effects, and reaction

Petroleum products are used across the globe by most industries and for different applications, where the requirement for the presence of petroleum materials on site is often imperative. This need stresses the necessity of moving petroleum substances using maritime ships or underwater pipelines internationally between different continents and countries. To move the oil from its source to the final consumer, up to 15 transfers are involved [1, 2]. Several reasons could cause the occurrence of oil spills: in addition to the intentional petroleum waste spill in seawater, transportation is vulnerable to involuntary oil spills from tanker collisions with rocky shoals, ship accidents, and pipeline ruptures [3]. On a global scale, the European Space Agency (ESA) has stated that the worldwide spillage exceeds 4.5 million tons of oil annually, where

45% of the amount is due to operative discharges from ships [4]. Accordingly, spills in seawater including light oil, gasoline, fuel, crude oil, and bulk oil have adverse long-term repercussions on the maritime environment. They happen on a global scale, influence maritime and human life, and have socioeconomic implications on communities [5, 6]. Depending on the ocean state and seawater currents, an oil spill can scatter over a very wide area in the open sea within only a few hours. Once the spill occurs, the oil quickly spreads to form a thick slick that becomes thinner over time when moving away from the source. This thin layer is called a "sheen" and has a rainbow-like appearance [1]. Therefore, oil spills can have varying thicknesses, ranging from less than a micrometer (um) for sheens to millimeters (mm) for thick slicks as documented in the Deep-Water Horizon case field samples [7, 8]. Based on the oil type and its thickness, a slick can evaporate quickly or can persist in the environment for a long time. For example, even though light oils are highly toxic, they evaporate quickly. But heavy oils, which are less toxic, persist in seawater for a much longer time and can get mixed with pebbles and sandy beaches where they remain for years [1]. Worldwide, fuels and crude oils account for 48% and 29% of the total oil spilled into the seawater, respectively, highlighting again how much the impact on the environment could be severe [9].

In the last decades, thousands of tons of oil were spilled worldwide. The quantity of oil spilled from tankers in the 1990s is estimated to be more than 1.13 million tons of oil, while 73% of the total amount occurred in 10 incidents only [10]. For example, during the Gulf War spill that occurred in 1991 on the sea island in Kuwait, around 800,000 tons of heavy crude oil were released from storage and ships [11]. Another huge spill occurred in 1992 in Fergana Valley in Uzbekistan due to an oil well blowout. During this accident, the estimation of the spilled oil is around 300,000 tons [2]. According to the international tanker owners' pollution federation (ITOPF), the estimated amount of oil spilled from tankers in the 2000s is 196,000 tons [10]. In 2010, while 4000 tons of diluted bitumen were spilled in Kalamazoo, Michigan, due to a pipeline rupture, an approximation of 500,000 tons of light crude oil were spilled in the same year in the Gulf of Mexico due to an oil well blowout [11]. Very recently in 2018, an oil spill of 113,000 tons size happened off Shanghai in China [10] in the Sanchi oil tanker collision.

While environmental rules, regulations, and strict operating procedures have been imposed to prevent oil spills, these measures cannot completely eliminate the risk [2]. Therefore, there is a need to set up oil spill contingency plans in countries where the chances of such disasters are high. These plans are usually prepared by the petroleum administration in close cooperation with ministries, governmental and non-governmental organizations as well as regional and international entities. They aim to protect human life, natural resources, and the economy by preserving the coastal and marine environment from any adverse effects an oil spill might pose. The main objective of such plans is to maintain the alertness and readiness of the involved entities to ensure a timely and effective response to oil spills and to prevent further pollution. In most scenarios, it requires the quick mobilization of aerial surveillance resources and consultation with international remote sensing institutions to provide oil spill modeling data or remote sensing images. Accordingly, to permit a swift and appropriate reaction and limit contamination, a need to establish an effective oil spill monitoring system presents itself in utmost urgency.

The contingency plan deals with various transformation processes including weathering, evaporation, and emulsification that occur when the oil is spilled on water. Also, the plan considers oil movements in seawater given the oil's physical characteristics and how it interacts with ocean waves. To contain the oil and prevent its spread over time, special equipment such as booms is used on-site. Afterward, devices such as skimmers and absorbing material such as sorbents are used to recover oil from the surface. Other alternatives are to use surface-washing and chemical dispersant agents to treat the oil or to have in-situ controlled burning as a cleanup technique [2]. All previous measures are very critical to making the contingency plan very effective. But a necessary factor for their usefulness is the pre-knowledge that is available about the spill. The more information is collected from the spill scene, the better the measures used for containment and cleaning-up operations are. The first and most critical information is the spill location. Other information, including the volume spilled or the type of the oil, is also necessary to define proper measures in the contingency plan. Since this information is not available at an early stage in many spills, surface techniques, as well as remote sensing techniques, are frequently used for the first assessment and evaluation.

Remote sensing is the field that combines science and technology to extract ("sense") information about an object or phenomenon at a distance ("remotely") by instrument-based techniques. A detailed description of the state-of-the-art remote sensing technologies for oil spill surveillance is listed in [1, 12–14]. The sensors used are usually mounted on aircraft or satellite platforms. They target the detection, thickness estimation, classification, or a combination of these functionalities. As an addition to the state-of-the-art systems, the aim of this chapter is to present a new advancement in oil spill monitoring using drone-based radar remote sensing, which complements currently available systems for better oil spill surveillance. Section 2 provides an overview of the oil spill monitoring system and gives the necessary background about the techniques used up to now for monitoring. It describes the pros and cons of the sensors (2.1) and platforms (2.2) used by state-of-the-art techniques to monitor oil spills. Furthermore, it describes the required information (2.3) and the main features (2.4) for good monitoring and an effective contingency plan. In addition, a system-level view of the new drone-based proposed solution is presented (2.5) and compared with recently developed state-of-the-art techniques in terms of their functionalities and the used sensors (2.6). Section 3 elaborates more on the new proposed approach for oil spill monitoring for oil spill detection, estimation, and classification. It describes in detail the system model (3.1), the new probabilistic detection algorithms (3.2), the new statistical thickness estimation algorithms (3.3), and the neural network regression algorithm for the oil type classification (3.4). The feature of onboard processing is elaborated more (3.5), and the system-level complementary solution is described (3.6). Section 4 concludes this chapter by summarizing the importance of the presented approach as a complementary solution to state-of--the-art techniques and highlights new aspects that should be considered in future developments.

2. Oil spill monitoring system

In this section, we present a brief overview of the state-of-the-art techniques used for oil spill monitoring in relation to the sensors used and to the platforms operated during the oil spill. Then, we motivate our new proposed solution by listing the system features, the required functionalities for an effective contingency plan, and how our solution differs from relevant state-of-the-art techniques. For more details about oil spill surveillance systems, the reader is encouraged to refer to [12–14].

2.1 Sensors

2.1.1 Visual sensors

Despite many shortcomings, passive sensors that operate in the visible region of the light are still used in oil spill remote sensing. The effect of some environmental conditions such as sun glint and wind sheen would lead to a misinterpretation by creating a resemblance to oil sheens, which is considered a limitation of such sensors [1]. Another drawback is that visual sensors cannot operate at night because they are using sunlight reflectance for operation. In addition, they require cloudless and clear weather requirements. Given the limitations of visual sensors for oil spill detection, and since they are not able to provide thickness information or oil classification, these sensors are not used alone for oil spill monitoring. For example, in [15], optical remote sensing images are combined with visible infrared imager radiometer suite (VIIRS) images in a semi-automatic fashion to extract oil slick features. The technique is tested in the North-West of the Gulf of Mexico. However, this method cannot be fully automated and requires human intervention to set a proper threshold for feature extraction. Optical sensors are rather used to document the spill and to provide a frame of reference for other sensors [16, 17].

2.1.2 Infrared sensors

Infrared passive sensors are relatively cheap remote-sensing technologies that can be used to detect oil spills [18]. The emissivity of the oil in the thermal infrared red region is lower than the emissivity of the water. This is how the thick oil could be distinguished from the background water by absorbing the infrared radiation from the sun and appearing as a hot spot compared with the cold background for the water [1]. An opposite phenomenon is observed during the night when the heat loss from the oil layer is faster compared with the water. This is the reason why they appear cooler at night [19]. However, false-positive results could be obtained by misinterpreting the thermal radiation from seaweeds. In addition, infrared sensors require the absence of cloud and heavy fog for good operation [1, 2, 20]. Infrared sensors can detect oil films with 10s–100 s um thickness. However, the brightness of the infrared sensing-based imagery does not vary with slick thicknesses in the mm range. Therefore, we cannot rely on infrared sensors to yield slick thickness measurements [13, 21, 22]. Chen et al. [23] shows that thermal infrared sensors that are mounted on helicopters can detect oil spills in the accident of the Dalian Xingang oil pipeline explosion in July 2010. Similarly, [24] shows that using MODIS thermal infrared data, the information obtained from the sea surface temperature identifies the oil film from seawater. This technique is applied to the Jiyyeh spill when Israel bombarded storage tanks in Lebanon during the war of 2006, where around 15,000 tons of heavy fuel oil spilled into the Mediterranean Sea [24, 25].

2.1.3 Ultraviolet sensors

Very thin oil films have a strong reflectance in the ultraviolet region compared with seawater. This allows the use of ultraviolet sensors for oil spill detection when the thickness is not greater than 10 um. Also, look-alikes such as sun glints, wind slicks, and biogenic material challenge ultraviolet sensors for oil spill detection [1, 3]. Huang et al. [26] proposed adaptive thresholding for chemical spill detection (not oil specifically) from ultraviolet images, which shows a distinction between the chemicals and the water background. Desbiens et al. [27] used ultraviolet range for remote detection of hydrocarbons such as benzene. Generally, fewer ultraviolet sensors are being used for oil spills in today's remote sensing because of the low relevance of thin slicks to oil spill cleanup [3, 13, 28].

2.1.4 Passive microwave radiometer sensors

Compared with water, the oil emits stronger microwave radiation and appears brighter in the background. Passive microwave radiometers [29–32] are used for both oil spill detection and thickness estimation [13]. The need to acquire knowledge about weather conditions, the low spatial resolution of this sensor, and the a-priori knowledge required about the oil characteristics all influence the microwave brightness and decrease the effectiveness of microwave radiometers for oil spill monitoring [1]. Furthermore, the main issue with this technology tends to be the cyclical relationship between the microwave brightness of the slick and its thickness. Currently, available models can only measure limited thickness ranges [14]. For example, using a multi-frequency passive microwave radiometer, the measured thickness range is limited between 0.1 and 1.5 mm as reported in [33], or the results were underestimating the real thickness values as in [30] where the calibration methodology and the selection of frequencies limited the measured thickness to a maximum of 1 mm. The only commercial tools currently available for measuring slick thickness are the Optimare 3–5-channel microwave instruments. They can provide thickness up to 3 mm only [13]. Given the requirement of a dedicated aircraft to mount this sensor, in addition to their high cost, it is complicated to put them into operation. Currently, the microwave sensor is not being used for oil detection and slick imaging [13].

2.1.5 Radar sensors

With the absence of oil slicks, a bright image is obtained by radar sensors for clean seawater. Once the oil is spilled into seawater, the ocean capillary waves are reduced, and radar reflections are decreasing. Dark spots are obtained in radar imaging. This allows for oil spill detection [34]. Synthetic aperture radar (SAR) and side-looking airborne radar (SLAR) are the two most common types of radar, which are used for oil spill remote sensing [35]. Imaging SAR systems [36–39] are off-nadir instruments whose backscattering over the ocean is primarily due to Bragg scattering at relevant incident angles. The synthetic aperture radar technique is highly prone to false targets, however, and is limited to a narrow range of wind speeds when small ocean waves do not yield a difference between the oiled area and the sea [2]. SAR techniques are not used for oil thickness estimations nor for oil classification. Being widely mounted on space-borne platforms, the radar is a very useful active sensor for a synoptic view of the oil spill over a wide scene [1].

2.2 Platforms: Airborne to satellites to complementary drones

Most recent techniques using one sensor, or a combination of sensors, are done remotely using airborne systems [40, 41] or satellites [42–47]. Radar satellites provide a selection of resolutions and polarizations [13]. Serious efforts have been made to replace airborne remote sensing with satellite remote sensing. However, satellites face the limitations of overpass frequency and low spatial resolution [12], and the long time required for processing the dataset, potentially disrupting oil spill contingency planning. This limitation has been improved using satellite constellations. A revisit time within a few hours can be provided by a larger number of SAR satellites. Airborne systems, despite their high cost due to aircraft dedication, can be used directly when needed for real-time dataset processing [1]. In addition, they provide flexibility in terms of deployment time and choices of sensors. Therefore, a combination of satellite and airborne sensors is used in many countries in northern Europe for oil spill surveillance. The strategic planning is based on satellite imagery that provides a synoptic view of the oil spill, whereas airborne sensors are used for short-term or tactical responses [9]. Contrarily to visible and radar sensors, due to the high atmospheric absorption and scattering, many sensors including the infrared and the fluorosensors are not suitable to be operated on a space-borne platform [34, 48].

Despite all the effort done using space-borne platforms, only 25% of the pollution cases are detected by satellite systems. For a quick response and rapid intervention, the European Maritime Safety Agency (EMSA) has proposed using drones as complementary systems in satellite maritime surveillance [49]. Aerial surveillance could be improved significantly through the introduction of drones because it is a quick assessment tool for oil spill accidents [50]. In addition, drone-based tools will be particularly valuable as it provides high spectral resolution, at a relatively low cost.

2.3 Required information

Once a spill occurs, the oil will spread quickly on the water surface to form oil layers such as slicks, films, and sheens. To alleviate the severity of oil spills and promptly react to such incidents, it is crucial to have oil-spill monitoring systems that enable an effective contingency plan [51]. A rapid response time and a quick intervention allow dictating the best actions to deal with oil spills. Therefore, monitoring systems must perform several functionalities and provide valuable information to contain the damage [52].

1. Functionality #1: Oil-spill detection

   The most important part during oil spills is to detect oil slicks. It is important to locate them and to determine how large the spread is. This necessary information allows oil spill mapping for both tactical and strategic countermeasures.

2. Functionality #2: Oil-thickness estimation

   The thickness distribution of spilled oil is another critical information for spill containment. Using the knowledge about the oil thickness, an estimation of the total volume spilled can be performed so that adequate tools are used in cleanup operations.

3. Functionality #3: Oil classification

   Classifying the oil type is also important during spill containment. Based on this information, the authorities estimate the environmental damage in the short- and long term to take appropriate response activities.

4. Functionality #4: Oil-spill tracking

Since the spill can spread quickly within a few hours, an effective contingency plan demands the ability to track the spill over time. Tracking provides timely and valuable information to anticipate possible damage scenarios, predict the trajectory using additional input from weather forecasts, and assist in cleanup operations.

2.4 Features for good monitoring

In addition to the four functionalities included in the previous section, it is also important to take the following system requirements [1] into consideration whenever an effective oil spill monitoring system is designed. This allows for selecting properly the sensor and the platform. In addition, suitable and effective algorithms could be developed.

- *Spatial resolution*: it is an important characteristic of the sensor that affects the accuracy in mapping oil slicks with accurate thickness measurements.

- *Quick response*: oil spill surveillance requires an acceptable time frame for collecting and processing the data, which stresses the need for real-time data availability.

- *Day-time conditions*: the operation of the sensor at any time during the day and night is essential for an effective surveillance system.

- *weather conditions*: it is important to limit the effects of weather conditions including rain and fog.

- *Large-scale view*: the sensor that captures a synoptic view of the area allows monitoring over a large spill spread.

- *Cost and size*: the acceptable size and cost of the sensor are critical to decreasing the overall cost of the system.

2.5 New proposed solution: Wide-band radar on drone platforms

Based on the previous criteria, visible and ultraviolet cannot work at night. Moreover, infrared sensors cannot provide estimations of thicknesses. Therefore, we select radar sensors for the proposed monitoring system because they operate during the day and night and under all weather conditions. Whenever mounted on satellites, radar is a very useful active sensor to detect oil over a large area. However, the synthetic aperture radar technique is limited to a narrow range of wind speeds. The ocean's slight surface roughness due to very low wind speeds (below 4 m/s) leads the backscattering to be dominated by the specular component, challenging SAR systems for oil spill detection [38]. Therefore, it would be advantageous to study the radar observations from nadir-looking systems (transmit and receive at zero angles from the normal to the ocean surface) since they cover scenarios that cannot be studied by SAR systems. Being largely independent of surface roughness, the returns from nadir (or near-nadir) systems will benefit from the dominance of the specular scattering and enable detection in very low wind conditions. Therefore, drone platforms are designated suitable for the proposed solution. The drone-based radar solution allows quick assessment of the area where the flag of possible spills is raised by witnesses. As a drawback, operating these platforms as nadir-looking systems decreases the surface of the scanned area viewed by the radar compared with that scanned by "side-looking" platforms. Also, using drones instead of satellites does not allow a synoptic view of the spill. But this can be compensated by using multiple drones at a time instead of scanning with a single drone. The parallelization in scanning can cover a large area at a critical time. Once a possible oil spill is announced, the drones can be directly used as tactical-response systems to scan the scene and report results. Afterward, whenever the spill is confirmed, the satellites can be used for strategic planning by providing a synoptic view of the spill area. Furthermore, scanning with drones provides a high spatial resolution compared with satellites and with a principal advantage of a relatively low cost compared with dedicated airborne detection systems.

Hence, we propose a new approach for oil spill monitoring with the following features, which are discussed in more detail in Section 3:

1. From a system-level perspective, we suggest the incorporation of both C-band and X-band using remote sensing nadir-looking wide-band radar sensors that can be implemented on drones as oil spill monitoring systems.

2. Our new approach targets the spills happening during calm and moderate ocean conditions, which are challenging for state-of-the-art SAR systems.

3. Compared with the state-of-the-art techniques, the newly proposed system implements maximum-Likelihood and machine learning statistical algorithms that can perform simultaneously in real time:

    a. functionality #1 by detecting thick oil slicks

    b. functionality #2 by estimating the thicknesses of detected slicks in the mm range

    c. functionality #3 by classifying the oil type

The proposed solution is not intended to replace state-of-the-art techniques but rather complements them by providing a complementary system to satellite SAR systems. During the early stages of a possible oil spill, drone systems act as small-scale tactical-response systems improving the large-scale surveillance obtained by satellite systems. Over the spill duration and based on satellite scans, the drones can track the spill using the high spatial resolution feature provided by the mounted wide-band radars.

2.6 Comparison to latest state-of-the-art techniques[1]

Several state-of-the-art techniques have been already included when we introduced the sensors (Section 2.1) and platforms (Section 2.2). To follow up on Section 2.5 where we describe the importance of the proposed approach—and for convenience

[1] It is important to keep in mind that in this section we are not including all state-of-the-art techniques used for oil spill monitoring irrespective of the sensors and platforms used.

—we only focus in this section on the recently developed techniques that target any of the three functionalities during oil spill monitoring (detection, estimation, and classification) using non-satellite-based approaches.

Several drone-based techniques are recently suggested for oil spill monitoring. Saleem et al. [53] presents thermal imaging methods and tools applied to simulated and lab-based experimental data to detect oil spills in seawater using drones. The simulation environment is made to be very similar to the Gulf of Mexico spill in 2010. The drone can be used in two operational modes: the first mode is when the spill incident is known, then the task is to locate it exactly. The second mode is when operating to look for potential spills and for nearby ships that caused them. After collecting thermal images using the thermal camera mounted on an unmanned aerial vehicle (UAV), the temperature profile of collected data is studied using image processing techniques. The proposed system shows the ability to draw a rectangular contour around the spill region but not the accurate contour. The thickness of the spill under test is not reported. Another drone-based solution equipped with a thermal camera is proposed in [54]. The system is suggested for industries, especially for the Bahrain Petroleum Company (Bapco), to inspect oil and gas leakages. It uses AI-based (decision tree (DT), random forest (RF), and support vector machine (SVM)) onboard processing to monitor oil pipelines for possible leakages and cracks. The training is done based on a real dataset of methane pipeline leakage detection provided by Bapco, Tatweer Petroleum, and the National Space Science Authority (NSSA). The proposed algorithms are accelerated on hardware using parallel processing to allow real-time alerting with less than 100 milliseconds delay. No results for oil spill leakage from pipelines are presented. The presented work is relying only on a single dataset while multiple datasets should be included. Oliveira et al. [55] also targets the first functionality of the monitoring system (detection) by presenting a UAV-based aerial solution to automatically identify the contour and localize oil spills. This approach is tested on simulated, experimental, and field data through simulated oil spills in the Leixoes Harbour and the Douro river in Porto in Portugal and in the Puerto de A Coruna in Spain. In [56], a deep-learning-based method is proposed to control marine oil pollution. It implements an SVM approach that processes visual camera images to identify oil-polluted areas. Afterward, it predicts the movement of the polluted area by calculating the optical flows. The algorithm is applied to the 2016 Tegent Cruises dataset in Shimen in northern Taiwan. The accuracy of the SVM classification varies between 85.71 and 99.95%. Another deep learning approach is suggested by [57]. It presents a CNN-based novel framework to detect small oil spills inside a port using a thermal infrared camera mounted on a drone. Three kinds of oil (HFO, DMA, and ULSFO) that are frequently used in the port of Antwerp are included during the experiment. A mean intersection over union (mIoU) of 89% is achieved. The proposed technique is functional within a 31.9 m by 42.1 m field of view and is useful for relatively small oil spills. Jiang et al. [58] uses hyperspectral images to feed a one-dimensional convolutional neural network to identify the type of oil spills. Using the adaptive long-term moment estimation (ALTME) optimizer, the oil spill spectral information is learned. The experiment is conducted 20 m away from the Yellow Sea Shore in China, in the pool of Qingdao Scientific Research Base. The technique achieves a detection accuracy of more than 98.09% in detecting different predefined classes of oil films of thicknesses between 1 and 3.5 mm. To summarize, [53–57] all target spill detection (functionality #1), whereas [58] also targets the thickness estimation (functionality #2) in predefined classes (between 1 and 3.5 mm) and the oil type identification (functionality #3). This reinforces the need for our new proposed system that targets the three different functionalities at the same time. Compared with the work presented in [58], the proposed new solution is using the wide-band radar sensor instead of a hyperspectral visual sensor (check differences in Section 2.1). Furthermore, the new approach uses maximum-likelihood algorithms on top of different machine learning algorithms to provide the results for monitoring. Also, the approach aims the thickness estimation up to 10 mm value, and it is feasible to be implemented on drones for onboard processing.

Other techniques, which are not drone-based, are also proposed in the literature for oil spill monitoring. Yin et al. [59] proposes an optical fiber surface plasmon resonance for oil spill detection and thickness estimation. At the water-oil and ail-oil interfaces, the sensor records an absolute sensitivity of 1.373%/mm and 2.742%/mm in the thickness ranges 0–5 mm and 0–10 mm, respectively. Although their approach can detect thicknesses in the 1–10 mm range, however, it requires in-situ sampling. Li et al. [60] presents multiple machine learning algorithms (RF, SVM, DNN, and DNN with differential pooling) that process images from the high-resolution hyperspectral sensor to identify the oil type based on the reflectance spectra. Four types of oil are tested in this work, including crude oil, diesel, lubricant, and heavy diesel. The oil thickness and the wind conditions are not provided as inputs to the machine learning models to test their accuracy with the minimal amount of available information. All tested models can differentiate between heavy oils. But for light oils, RF fails to do the correct classification, while the neural network models provide better classification than the SVM. This technique is not remote-based and should be used after the detection of oil spills. Dala et al. [61] develops a novel microwave oil spill sensor to determine the thickness in the range of 10s mm using an ultra-wideband radar operating between 0.3 and 3000 MHz. The performance of the system is tested under static experimental conditions. When deploying the system in the sea, the impact of the waves should be considered.

A comparison between the functionalities provided by the developed techniques above and our new approach is presented in **Table 1**.

3. Drone-based radar approach

In this section, we describe in detail the proposed approach for oil spill monitoring for oil spill detection, estimation, and classification.

3.1 System model

3.1.1 Reflection coefficient for multi-layer structure

To physically model the oil slicks on top of the sea surface, we consider a multi-layer structure: air, oil, and seawater. We study the reflection of the electromagnetic waves from the sea layer covered by an oil layer, which is modeled as a thick slick with a thickness of up to 10 mm [13]. **Figure 1** shows such a simulated oil spill. When the ocean is calm and the wind speed is low, such a thick layer significantly reduces the surface roughness [37, 62], dampens capillary and short gravity waves, and calms the smaller waves on the sea [63]. As a result, within the radar cross section, the incident electromagnetic waves are reflected along the specular component due to the smooth surface. By operating the radar on a nadir-looking system, the reflected power is fully reflected for normal incident waves, without losses due to off-nadir backscattering.

Sensor(s)	Detection	Thickness Estimation	Classification	Notes	Reference
Thermal camera	Yes	No	No	Image processing	[53]
Thermal camera	Yes	No	No	AI-based techniques (DT, SVM, RF)	[54]
Visual/Thermal cameras, Lidar	Yes	No	No		[55]
Visual camera	Yes	No	No	SVM Prediction of track	[56]
Thermal infrared	Yes	No	No		[57]
Hyperspectral sensor	Yes	Yes [1–3.5 mm]	Yes	Post Processing	[58]
Fiber-Optic Surface Plasmon Resonance	Yes	Yes [0–10 mm]	No	Not drone-based, using in-situ sampling	[59]
Hyperspectral sensor	No	No	Yes	Not drone-based, using RF, SVM, and DNN	[60]
Ultra-wideband radar	No	Yes [10s of mm]	No	Not drone-based, under static conditions	[61]
Wideband Radar	Yes	Yes [0–10 mm]	Yes	Onboard processing	Proposed solution

Table 1.
Non-satellite-based state-of-the-art techniques for oil spill monitoring.

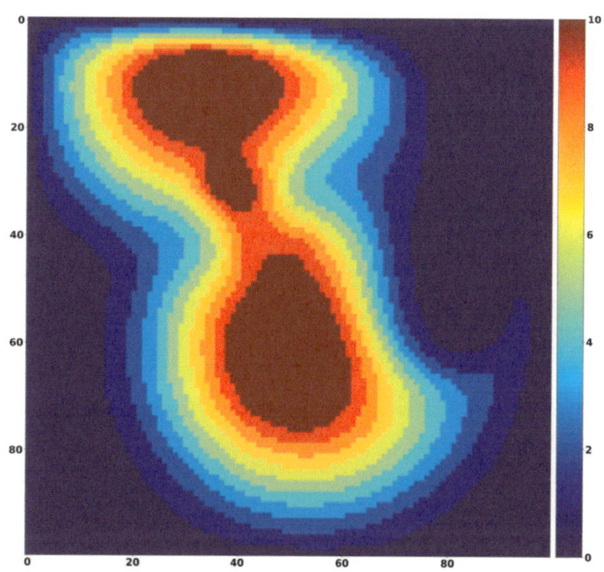

Figure 1.
Simulated oil spill with thickness range from 0 to 10 mm.

We define the electrical properties (relative dielectric constants) for the air, oil, and seawater to be respectively ε_1, ε_2, and ε_3. The different media are assumed to be non-magnetic. If radar reflections from the seafloor are neglected, we can calculate the field reflection coefficients for the layers at the boundaries where interaction with electromagnetic waves occurs. They are denoted for the first interface (between air and oil) and the second interface (between oil and water) by ρ_{12} and ρ_{23}, respectively. Hence, the power reflection coefficient for the three-layer structure, also denoted by the reflectivity R, is derived using the Transfer Matrix Approach [64, 65] as:

$$R = \left| \frac{\rho_{12} e^{j\delta} + \rho_{23} e^{-j\delta}}{e^{j\delta} + \rho_{12}\rho_{23} e^{-j\delta}} \right|^2 \tag{1}$$

δ is the propagation constant that is dependent on the oil's relative permittivity (ε_2), the frequency of the electromagnetic wave (the wavelength λ), and the thickness of the oil layer (d). It is given by:

$$\delta = \frac{2\pi \sqrt{\varepsilon_2}\, d}{\lambda} \tag{2}$$

3.1.2 Smooth and rough surfaces

Several statistical attributes can be calculated for a random surface [66]. The surface height measurements may be described using standard statistical parameters, namely:

- the height standard deviation (s), which is also called the rms-height
- the surface correlation length that measures the degree of correlation between the surface at different locations

Based on the surface statistics with respect to the electromagnetic wavelength, we can differentiate between two types of surfaces: a perfectly *smooth* (flat) or *rough* surface. If the correlation length of the ocean waves is large and the rms-height of capillary waves is very small, as it is for calm ocean conditions, then the surface is considered smooth. In this case, radar reflections are along the specular direction and can be calculated according to Eq. (1). Otherwise, a non-coherent component is introduced in the scattering pattern along all directions other than the specular, whenever the surface is rough. For this, with θ denoting the incident angle of the electromagnetic waves to the interfaces, the reflectivity in the specular direction is called the coherent reflectivity $R_{coherent}$, and it is given by:

$$R_{coherent} = R.e^{-4\left(\frac{2\pi\, s\, \cos(\theta)}{\lambda}\right)^2} \tag{3}$$

3.1.3 Reflectivity behavior

For our approach, we suggest the incorporation of multiple bands using remote sensing nadir-looking wide-band radar sensors. Since frequencies from the L-band (1–2 GHz) will not change the reflectivity value over the 1–10 mm thickness range by more than 1 dB, they are not considered to be relevant for the proposed approach. Instead, we will use frequencies from the C-band (4–8 GHz) and X-band (8–12 GHz).

To better understand the reflectivity behavior at each wave frequency, **Figure 2** shows the reflectivity value defined in Eq. (1) versus the oil thickness, which is varied from 1 mm to 10 mm where different subplots correspond to different scanning frequencies at increments of 1 GHz from 4 to 12 GHz [67]. The relative dielectric constant of the air and thick oil is 1 and 3, respectively. We neglect the imaginary part of the relative permittivity of the oil, which is of order 0.01 [68]. The relative dielectric constant of seawater depends on the water temperature (assumed to be 20 degrees), the water salinity (assumed to be 35 ppt), and the frequency of the radar signal [66]. For ease of reference, we also mark the mean reflectivity value in the case of no oil spill as a constant straight line per subplot.

For instance, at 4 GHz, the curve of the reflectivity is monotonically decreasing slowly with the thickness. It has a very small slope at small thickness values (0–3 mm), but this slope increases with the increase of the oil slick thickness (3–7 mm). At some thicknesses, any error in the power reflectivity measurements at 4 GHz would mislead the oil detection due to the very small variation from the water reflectivities. In addition, for small thickness values, the difference between reflectivity values is very small, so the estimation could go easily wrong. A different pattern is observed at higher frequencies (7–12 GHz). The reflectivity curve admits a steeper slope for small values of thicknesses, which improves the detection of oil slick thickness. However, due to the appearing cyclic behavior, many thickness values give the same reflectivity value leading to false interpretations for detection or estimation. Therefore, it is important to use more than one frequency to improve the detection or estimation. To see the effect of the relative permittivity of the oil on the reflectivity, **Figure 3** studies the variation of this parameter evaluated at two frequencies: 4 and 12 GHz [69]. For the same frequency, the effect of the variation in the relative permittivity on the reflectivity value is dependent on the oil thickness. Also, by

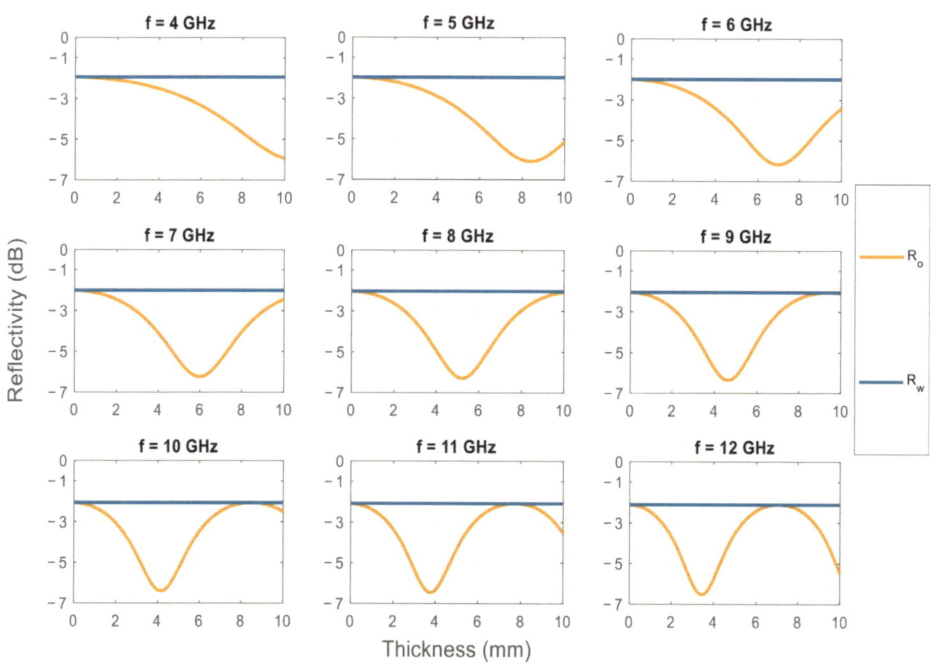

Figure 2.
Reflectivity (in dB) versus oil slick thickness (in mm) at different scanning frequencies. Retrieved from [67].

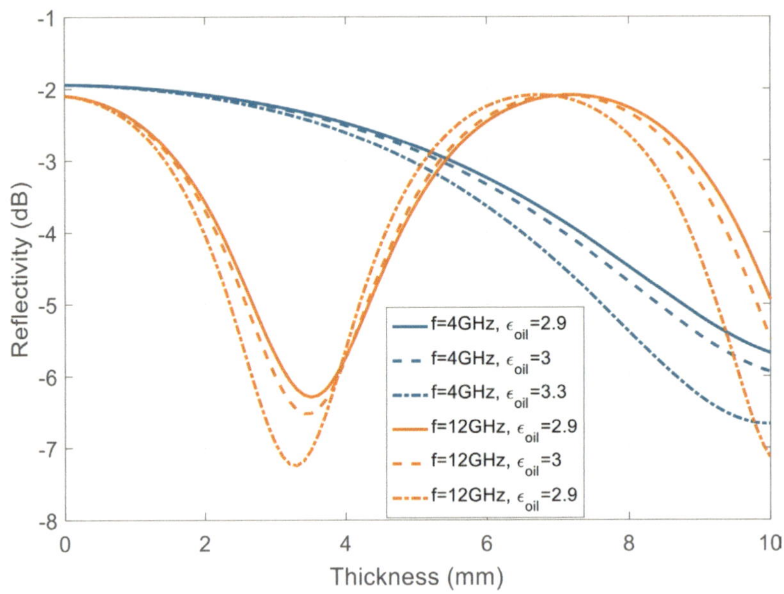

Figure 3.
Reflectivity R (in dB) versus oil thickness (in mm) at different frequencies (4 GHz, and 12 GHz) and different oil dielectric constants. Retrieved from [69].

looking at the reflectivity plots at the same frequency, they look very similar over the full thickness range even if the relative permittivity of the oil slick is changing. However, the reflectivity behavior varies a lot from one frequency to another.

3.2 Detection

To perform the detection functionality of the monitoring system, we are proposing in this section different detection algorithms that use the statistical characterization of the reflectivity values and their distribution under different oil thicknesses to obtain a final decision on whether oil exists or not. Any previous knowledge about the existence or absence of oil in the surface scanned should be taken into consideration to weigh the probability of the decision in the detector block. Nevertheless, without any previous knowledge about the spill situation, the detector decision will be totally based on the statistics of the measured power reflection coefficients ratio [69].

Let "o, w" be the events indicating the presence of the oil slick and the water, respectively. Let R be the event representing the measurement of the reflectivity value (s). R could represent one or more reflectivity values that are measured at the same or at different frequencies. The reflectivity values are assumed to be independent events, i.e., they are uncorrelated in the time domain at multiple observations, and in the frequency domain at multiple frequency measurements. After some derivation steps shown in [69], the detector algorithm evaluates the following ratio:

$$\frac{\Pr(o|R)}{\Pr(w|R)} = \frac{\Pr(R|o)}{\Pr(R|w)} \underset{<_w}{\overset{>^o}{}} 1 \tag{4}$$

With this ratio, the detector compares the probability of getting an oil or water event given that radar reflectivities are measured. If the ratio gives a result greater than unity, the decision indicates the oil's existence. Otherwise, the decision indicates that the seawater surface is clean. We note that the probability of obtaining a measured reflectivity value given that the oil or water exists is evaluated using the corresponding probability density function (pdf). For this, we propose several algorithms for different detectors:

- *Single observation at multiple frequencies* [70]: a wide-band radar is used to scan a target scene once, and K measurements of reflectivity are sampled at K different frequencies.
- *Multiple observations at single frequency* [71]: a narrow-band radar scans a target scene multiple times. M measurements of reflectivity are made at the same frequency.
- *Multiple observations at multiple frequencies* [69]: a wide-band receiver scans a target scene M time, and the reflectivity is sampled per scan at K different frequencies. M*K measurements of reflectivity are collected in total.
- *Joint-probability density function* [72]: the thickness of the oil slick is modeled as a random variable (RV) with an estimate of its probability density function. The algorithm is using the joint-pdf to calculate the probability of each event. Also, for this, a wide-band receiver scans a target scene M times, and the reflectivity is sampled per scan at K different frequencies.

Figure 4 shows the performance of dual-frequency detectors. For clarity, we also show the performance of the detectors running each of the two scanning frequencies separately. Evidently, the dual-frequency detectors outperform the single-frequency

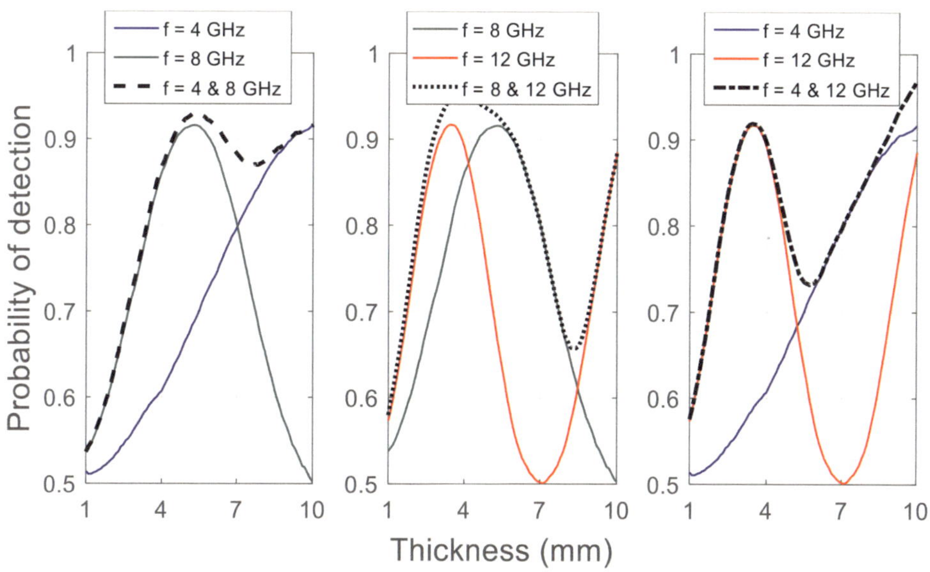

Figure 4.
Comparison between the probability of detection and the oil thickness (in mm) for different detector algorithms: Single-frequency detectors at 4 GHz, 8 GHz, and 12 GHz, and dual-frequency detectors using combinations of these frequencies.

detectors. At thickness values where the accuracy of a single-frequency detector is low, the dual-frequency detector gives more weight to the decision of the pair detector. For instance, a 12 GHz detector performs poorly for d = 7.2 mm because it has the same reflectivity value as the water (refer to **Figure 2**). However, the (4 GHz, 12GHz) detector optimally tracks the performance of a 4 GHz detector.

Although dual-frequency detectors are better than single-frequency detectors, there are still some thickness ranges where the probability of detection is affected by the ambiguity points. To boost the performance of the detection, we increase the number of frequencies K and observations M. Results in **Figure 5** show that a higher probability of detection is further obtained over the full range of thicknesses. Using three frequencies (4 GHz, 8 GHz, 12 GHz) and two scans (M = 2) of the target scene, the detection accuracy is above 90% for all thickness values above 3 mm. Additional scans would further be required to improve the probability of detection of thin oil slicks. Fortunately, performance requirements are less strict for small thickness values. By comparing the performance of the tri-frequency detector with single observation to the dual-frequency detectors with double observations, we notice that the first detector is better since it provides a very high percentage of correctness at the large thickness values (i.e., cases where oil detection is crucial) and very good behavior at small thickness values, which are more challenging for oil detection but less severe in false alarm scenarios. Additionally, the performance of the detectors is studied under a blind detection scenario in which the scanning of the scene is done without any knowledge about the exact or the estimated thicknesses of spilled oil. This type of detection is based on the joint-pdf that takes into consideration all the possible thicknesses with their probabilities in order to weigh the final decision. The distribution of oil thickness is assumed to be uniform over the range (0–10 mm). Results show the same behavior as detailed in [72]. The detector can provide correct decisions in different thickness ranges at different frequencies. The probability of error can be decreased by using dual- and tri-frequency detectors. Additional error can be

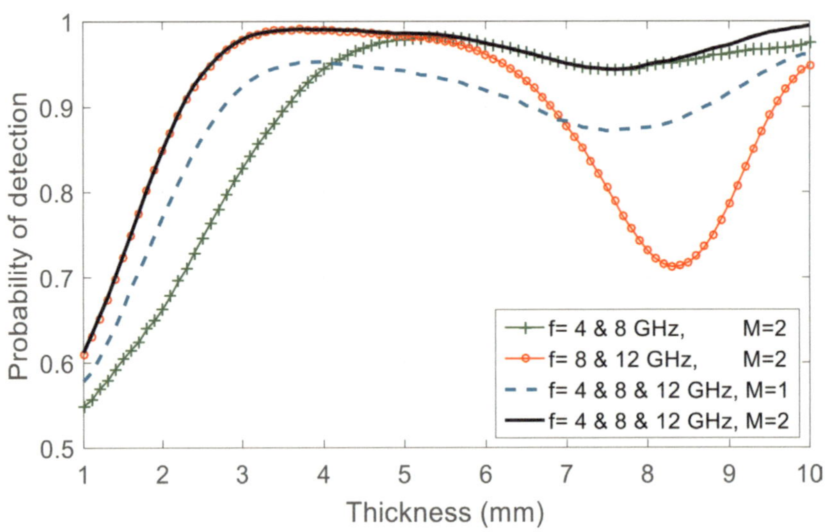

Figure 5.
Comparison between the probability of detection and the oil thickness (in mm) for different detector algorithms: Dual-frequency detectors and multi-frequency detectors, using single (M = 1) and double observations (M = 2).

compensated by using more scans. However, in the range where the oil thickness is small (< 2 mm), the probability of detection does not exceed 65% indicating low certainty.

Previous results are obtained under the assumption that the surface is smooth due to very low wind speeds and calm ocean conditions. For higher wind speeds and rms-height of the seawater, the surface roughness increases. We present the performance of the detectors for different roughness scenarios in detail in [69]. Results show that the performance overall is reduced.

3.3 Estimation

In this section, we complement the detection of oil spills (functionality #1) with a new approach targeting the estimation of the oil slick thickness (functionality #2). We present Maximum Likelihood single-, dual-, and multi-frequency estimators. The reflectivity values evaluated at one, two, or multiple frequencies over all possible thickness values form the constellation set that is used for the estimation. The estimator uses the Minimum-Euclidean distance algorithm, in predefined 1-D, 2-D, or K-D constellation sets, on radar reflectivities to estimate the thickness of the oil slick. Every constellation set is divided into different mapping regions that are bounded optimally between all the theoretical reflectivity values constituting the constellation. Any reflectivity value R obtained in one region is mapped by the estimator to the thickness that provides theoretically the nearest reflectivity value.

The derived algorithms are the following:

- *1-D estimator* [73]: It is a single-frequency estimator. The probability of error is minimized by choosing the thickness, which minimizes the Minimum-Euclidean distance between the measured reflectivity R and all possible calculated reflectivities constituting the constellation set. For the single-frequency estimator, the constellation set is in a single dimension.

- *2-D estimator* [73]: The dual-frequency estimator uses two reflectivity values in the constellation set spanning two dimensions.

- *2-D estimator iterative approach with multiple observations* [74]: The presented algorithm uses optimized predefined 2-D constellation sets by utilizing the best pair of frequencies for each possible thickness value. Then, it processes sequentially the separate estimations done to optimize the estimation procedure.

- *K-D estimator* [67]: Using K-frequencies in the estimation, the constellation set will span K dimensions. When using M observations that are uncorrelated in time, we do the averaging of the measured reflectivity values in K-dimension at each frequency of measurement.

- *SVM approach* [75]: A support vector regression that is following a supervised learning approach and is trained on reflectivities calculated for three radio-wave frequencies is proposed to predict oil thicknesses between 1 and 10 mm.

- *ANN approach* [76]: An artificial neural network that processes reflectivities evaluated at nine radio-wave frequencies is proposed to predict oil thicknesses between 1 and 10 mm.

Using a single-frequency estimator in C-band (4–8 GHz), the mapping regions at low thickness values will be close to each other as shown in detail in [67, 73]. A small noise power may shift the reflectivity values from one region to another. If so, this will be considered an error in the estimation process. At higher frequency values in X-band (8–12 GHz), many thickness values will give the same reflectivity due to its cyclic behavior. Therefore, even in an ideal case when no noise is introduced during the measurements, the estimator is not capable to provide a correct estimation. *This highlights the need to use the combination of reflectivity measurements (2-D or 3-D estimators) to achieve accurate thickness estimation.* Using dual- and tri-frequency estimators, one question will rise: which combination of frequencies is to be used? The best frequency pairs and triads, for each possible thickness between 1 and 10 mm, are derived in [67, 74] respectively. Daou et al. [74] also proposes an advanced iterative procedure to use the 2D estimator for accurate and reliable thickness estimations.

Figure 6 shows the histograms of the thicknesses estimated by K-D estimators when the actual thickness d is 5 mm with single and multiple (M = 3) scans. The single-frequency (1-D) estimator uses f_1 = 4 GHz, and f_2 = 12 GHz. The dual-frequency (2-D) estimator uses both f_1 and f_2. The 3-D estimator uses an additional frequency f_3 = 7 GHz. The 4-D estimator uses f_1, f_2, f_3 in addition to f_4 = 10 GHz. The 2-D estimator with three scans still provides thickness values from all possible values. The 4-D estimator with three scans decreases the error in the estimation to 4 mm with a probability of correctness of 80%. The percentage error obtained by all estimators under different scenarios is presented in **Table 2**. To check the distribution of the histograms of errors in estimation, refer to Appendix A in [67].

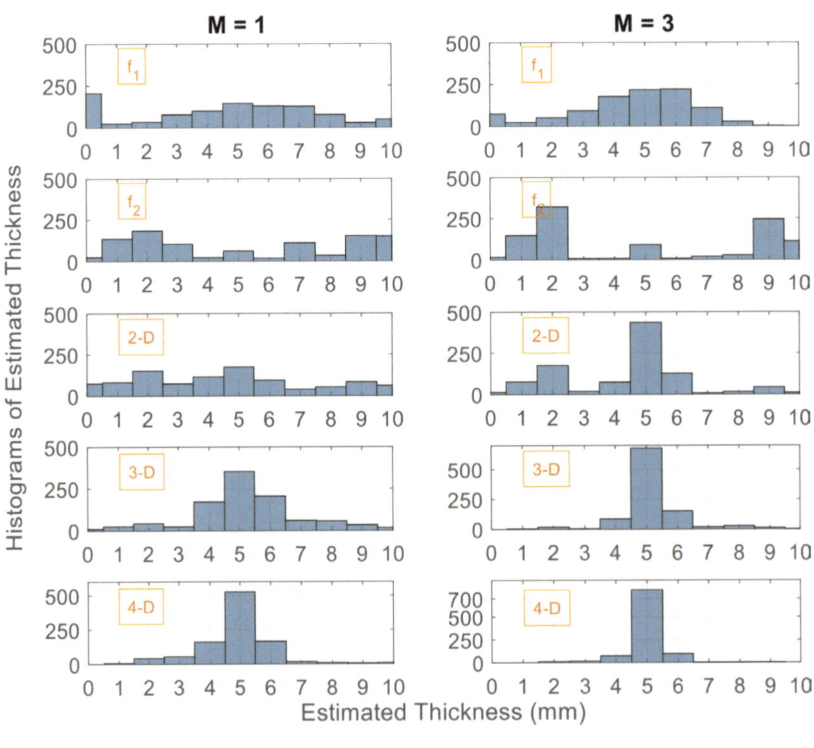

Figure 6.
Histograms of the thicknesses estimated by 1-D, 2-D, 3-D, and 4-D estimators with single and multiple scans M = 3. The actual thickness d is 5 mm. Retrieved from [67].

Probability of error (%)	M	f_1	f_2	2-D	3-D	4-D
d = 10 mm	1	46	78	35	32	30
	3	42	63	14	13	11
d = 5 mm	1	85	94	83	65	47
	3	77	91	58	33	20
d = 1 mm	1	97	85	83	77	72
	3	94	75	65	56	49
	10	90	58	36	32	26
	50	79	41	10	9	6

Table 2.
Percentage error obtained by different estimators for different slick thicknesses. Adapted from [67].

Although some estimators provide a good probability of correctness, it is not sufficient to have a definite decision. We should rather look at the error distribution in the estimation to ensure that the estimations are reliable. Therefore, we should always increase the number of frequencies or scans done to make sure that good and enough estimation information is provided.

We developed an experimental setup to collect radar reflectivity measurements from an oil-spill lab experiment with calm water surface conditions for very low wind speeds with no wave action. The experimental setup includes details about the system model with the multi-layer structure, the radar calibration technique, and other setup parameters [77, 78]. We applied the estimation algorithm to these experimental values. **Figure 7** shows the histograms of estimated thicknesses by 1D, 2D, 3D, and 4D estimators based on experimental reflectivity values with a single scan (M = 1), when the actual thickness is 3 mm. As expected, 1D estimators using a single frequency are not providing very good results. Using f1, the highest estimations are for 4 mm, the probability of error is 72%, and the maximum deviation from the correct value is 7 mm. Using f2, the deviation is decreased to 1 mm with a probability of error of 47%. However, single-frequency estimators are never used alone, therefore, including f2 in the 2D estimator again increases the probability of error to 68% but provides the advantage of decreasing the maximum deviation to 1 mm. For the higher-order 3D estimator, the probability of error decreases to 28% with the same maximum deviation. For the 4D estimator, the performance is further improved, and the probability of error is reduced to 6%. Thus, the results shown validate the proposed higher-order estimators on in-lab experimental data.

Another proposed estimation algorithm is based on a machine learning approach [75]. For this, a support vector regression (SVR) model that is following a supervised learning approach is trained on reflectivities calculated for three radio-wave frequencies to predict oil thicknesses between 1 and 10 mm. The input features to the SVR model are the power reflection coefficients (reflectivities) evaluated at f_1 = 4.39 GHz, f_2 = 6.98 GHz, and f_3 = 9.07 GHz. After training the model using 80% of the simulated data, it is validated on the remaining 20% of the simulated and on in-lab experimental data. The model yielded an R^2 score of 0.992 on the simulated data, which is very close to 1 indicating that the regression predictions are very close to the actual oil thicknesses. The model is supposed to predict a continuous value of the thickness as the predicted output. But by rounding the estimated thickness to the closest integer, we

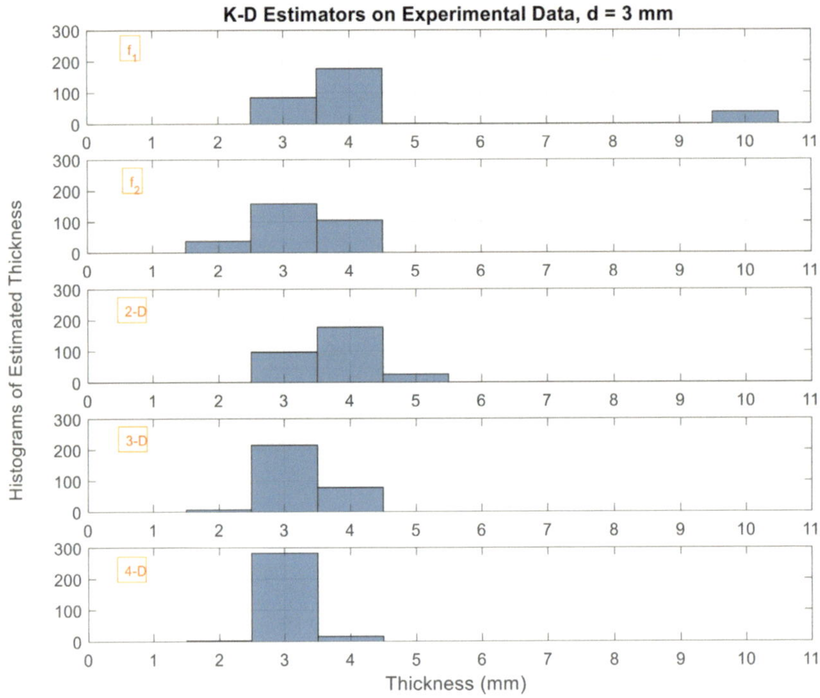

Figure 7.
Histograms of the thicknesses estimated by 1D, 2D, 3D, and 4D estimators based on experimental data with single scan M = 1. The actual thickness is 3 mm. Retrieved from [67].

get the following percentages of estimating the correct value in the confusion matrix: [62.4, 95.5, 99.7, 97.7, 97.7, 99.6, 99.4, 95.4, 80.5, 89.6%] for the thicknesses [1, 2, 3, 4, 5, 6, 7, 8, 9, 10 mm] respectively. The R^2 score obtained on the experimental data is 0.86 indicating that the experimental thickness values and the predicted thicknesses are close even though the model is only trained on simulated data and is not exposed to any experimental reflectivity. For further validation, even though the model is trained on calm surface data, it is tested with rough surfaces in two scenarios for slick thicknesses: 5 mm and 8 mm. The model's predictions are mostly concentrated at 6 mm for the 5 mm spill. Similarly, for the 8 mm spill, the predictions are mostly at 7 and 8 mm. This shows that using a machine learning algorithm for oil thickness estimation is an attractive approach. Another algorithm is developed in [76], but for convenience, we will discuss it in the following section. This concept can be further developed by using larger and more complex machine learning models and more diversified training data to achieve better thickness estimations in varied environments.

3.4 Classification

An important task to perform during oil spills is to specify the oil type to predict the environmental damage to maritime life. The oil type classification could be qualitatively similar to what is proposed by [60, 79–81]. Another way for classification is by analyzing the physical characteristics of the oil material, namely the relative dielectric constant, as we proposed in [76]. In this work, we use an artificial neural network (ANN)-based model to estimate the relative permittivity of oil slicks for different oil-types classification. In addition, the model can predict the thicknesses of such oil slicks at the same time. The input features to the model are nine radar reflectivity values measured simultaneously and selected uniformly from C- (4–8 GHz) and X- (8–12 GHz) bands. The ANN-model post-process these measurements to extract the implicit information about the thickness and the relative permittivity of the thick oil layer covering the sea surface, even though the reflectivities and the estimated parameters have a highly nonlinear relationship. To further improve the model's accuracy, we also incorporate multiple observations to boost the estimator's performance. Results show that by jointly analyzing the reflectivity behavior at multiple frequencies, the model explores their dependence over the full plausible range of thicknesses. The predicted thicknesses are accurate to +/− 0.5 mm in most cases. For example, testing the model at 5.5 mm would still lead to estimations between 5 and 6.1 mm. Similarly, relative permittivity values 2.9, 3, and 3.1 are approximated accurately where the shift in estimations' mean is smaller than 0.05. The error in estimations for the remaining values 2.8, 3.2, and 3.3 is higher. To test the performance of the model, we simulate an oil spill scenario when the thickness of the oil slick that is close to the source is 10 mm, and it decreases gradually going away from the source to 1 mm. The trained ANN model shows that it can perfectly reconstruct the spill environment with accurate estimations of the thicknesses and the relative permittivity, where the error of the average estimate for the latter over the full map is 3% only. To further validate the performance of the model, it is applied to in-lab experimental data when the actual thickness is 7 mm. After collecting 13 measurements, the estimated thickness is 7.65 mm compared with 8.2 with a single scan measurement. Unfortunately, we are not able to compare the estimated permittivity with the actual value since the latter has not been measured during the in-lab experimental procedure.

3.5 Onboard processing

Our proposed approach targets the implementation of the monitoring functionalities on board, instead of collecting measurements from the site and post-processing them offline afterward. This will ease the monitoring and most importantly will allow the quick assessment of the spill area, to take the best measures and actions toward the possible damage. For this reason, it is very important to check the feasibility of implementing the proposed approach onboard, especially since the platform that we are suggesting is the drone, which is a battery-powered device where the energy and power consumption are of high importance. We show the feasibility of our approach for onboard processing by selecting one of the previously described algorithms and implementing it on a hardware platform, namely the Pynq Z1 Field Programmable Gate Array (FPGA). While machine learning algorithms are often considered high-power consuming and introduce a high complexity in terms of memory footprint and computational requirements, we select the ANN model that performs the estimation and classification functionalities for the implementation. It is also good to note that including "0 mm" as one class for the regression estimation is equivalent to including the detection functionality. Therefore, we believe that the implementation complexity of this model is very significant since the latter can perform all the main required functionalities for oil-spill monitoring at the same time. Results show that, based on the architecture used during the implementation, only 1.13–28.37% of the available resources are utilized by the network. Similarly, the power consumption varies between 34 and 133 mW, which is negligible compared with the power drawn by drones in the order of 10th of watts. This verifies the feasibility of our ANN-based approach and demonstrates the suitability for a practical scenario.

3.6 The proposed system: Complementary solution

For completeness of the system-level analysis, we suggest having a complete system composed of four subsystems as shown in **Figure 8**:

1. Drone: It is the part used for tactical response. Whenever there is a flag raised for possible oil spills (by witnesses or due to accidents), multiple drones that mount nadir-looking wide-band radar sensors are sent to place. Measurements at different frequencies are collected. Based on the weather conditions and the ocean's waves, multiple scans could be used to improve the accuracy of the results. The measurements are then processed onboard using the proposed algorithms to provide an initial assessment with respect to the detection, estimation, and classification functionalities.

    a. The detection algorithms proposed in Section 3.2 are applied. If the probability of detection exceeds the threshold set by the persons in charge, then there is a need to prepare to start the contingency plan.

    b. The estimation algorithms proposed in Section 3.3 are applied. Thickness values greater than 1 mm indicate the need for quick intervention because the oil slick will persist for a long period of time.

    c. The classification algorithm in Section 3.4 is applied to have an indication about the oil type.

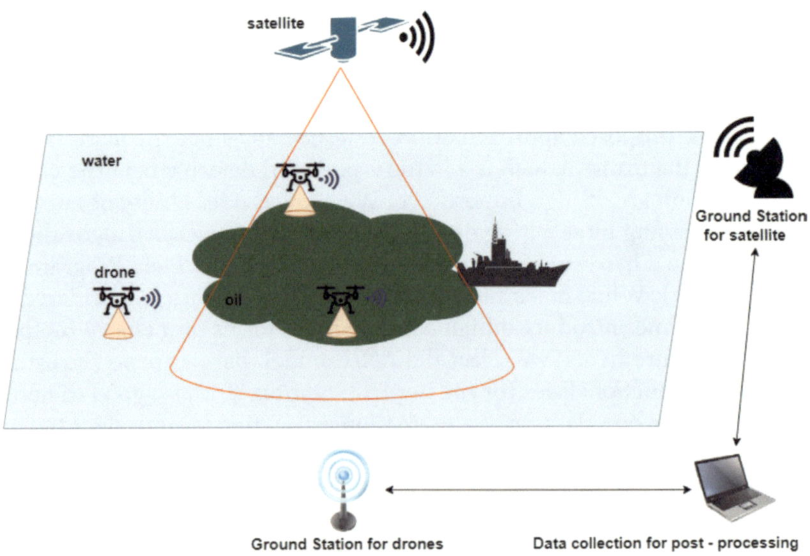

Figure 8.
System-level view of the oil spill monitoring system.

d. The results are sent to a ground station to combine results from different drones and locations on the corresponding map.

2. Satellite: It is the part that can provide the synoptic view of the oil spill scene. After the confirmation of the spill based on the initial assessment, the satellites are used to provide the synoptic view and to get a better estimate of how the spill is displaced over time. The satellites are supposed to transmit SAR images to corresponding ground stations.

3. Data collection center: It is the part responsible for data collection, analysis, and making decisions. SAR images are analyzed and compared with results received from drones. Corresponding measures are taken, and the oil spill contingency plan is launched. Oil spills are then tracked over time.

4. Conclusion

This work shows that by processing radar power reflectivity values, taken from nadir-looking systems under weather conditions suitable for cleaning operations, thick oil slick thicknesses up to 10 mm can be detected, estimated, and classified. The accuracy of each function is dependent on the selected statistical algorithm. This approach is novel since it provides most of the information that is required for an effective contingency plan at the same time during the early stage of the spill. In addition, this novel approach being implemented on drone platforms is a suitable system for tactical responses needed during contingency plans. Moreover, the new approach aims to complement state-of-the-art techniques, such as satellite-based SAR, by covering calm ocean conditions and low wind speeds scenarios that challenge other techniques for oil spill monitoring. Maximum-likelihood and machine learning statistical algorithms could be used directly to quickly assess the scene. The ANN model shows very low complexity, low power consumption, and high accuracy. This demonstrates the feasibility to apply the approach for onboard processing. Despite all the advancement that is presented, there are still open questions, tasks, and challenges to consider in this field for utilized practical solutions. A further investigation on the algorithmic level is still required to make sure that the performance will not decrease given the variable dynamics that could be introduced in the physical environment. Also, the complexity of all proposed approaches should be studied to check the feasibility of implementing them on hardware platforms for onboard processing. Despite the need for detection, thickness estimation, and classification, tracking the spill is also very important on a small scale, especially for moderate wind speeds. It would be useful to add this functionality as a capability of the drone-based solution by considering the weather conditions and the ocean state to track the spill over time. Finally, the proposed work is a proof of concept and helps take the oil-spill-related research work one step forward toward the development of operational tools for oil-spill intervention.

Conflict of interest

The authors declare no conflict of interest.

Notes/thanks/other declarations

The authors would like to thank the remote sensing center (RSC) at the national council for scientific research in Lebanon (CNRS-L) for its support. Also, special thanks to C.S.T.B, the electromagnetic studies department, at PHELINE in Grenoble, France, for the help in conducting the oil spill in-lab experiments and radar measurements.

Author details

Bilal Hammoud* and Norbert Wehn
Microelectronic Systems Design Research Group (EMS), Technical University of Kaiserslautern, Kaiserslautern, Germany

*Address all correspondence to: hammoud@eit.uni-kl.de

References

[1] Jha MN, Levy J, Gao Y. Advances in remote sensing for oil spill disaster management: State-of-the-art sensors technology for oil spill surveillance. Sensors. 2008;**8**(1):236-255

[2] Fingas M. The Basics of Oil Spill Cleanup. Boca Raton, FL, USA: CRC press; 2002

[3] Grüner K, Reuter R, Smid H. A new sensor system for airborne measurements of maritime pollution and of hydrographic parameters. GeoJournal. 1991;**24**(1):103-117

[4] Oil Pollution Monitoring. Remote Sensing Exploitation Division. ESRIN—European Space Agency (ESA). p. 2. Available online: http://www.esa.int/esapub/br/br128/br128_1.pdf. [Accessed: December 18, 2018]

[5] Ipingbemi O. Socio-economic implications and environmental effects of oil spillage in some communities in the Niger delta. Journal of Integrative Environmental Sciences. 2009;**6**(1):7-23

[6] Sandifer PA, Ferguson A, Finucane ML, Partyka M, Solo-Gabriele HM, Walker AH, et al. Human health and socioeconomic effects of the Deepwater horizon oil spill in the Gulf of Mexico. Oceanography. 2021;**34**(1): 174-191

[7] Svejkovsky J, Hess M, Muskat J, Nedwed TJ, McCall J, Garcia O. Characterization of surface oil thickness distribution patterns observed during the Deepwater horizon (MC-252) oil spill with aerial and satellite remote sensing. Marine Pollution Bulletin. 2016;**110**(1): 162-176

[8] Belore R, Trudel K, Morrison J. Weathering, emulsification, and chemical dispersibility of Mississippi canyon 252 crude oil: Field and laboratory studies. In: International Oil Spill Conference Proceedings (IOSC), Tampa, FL, USA. Washington, DC, USA: American Petroleum Institute; 2011

[9] Brekke C, Solberg AH. Oil spill detection by satellite remote sensing. Remote sensing of environment. 2005; **95**(1):1-3

[10] ITOPF, Oil Tanker Spill Statistics 2021, Available online: https://www.itopf.org/knowledge-resources/data-statistics/statistics/. [Accessed: August 1, 2022]

[11] Fingas M. Introduction to oil spills and their clean-up. In: Petrodiesel Fuels. Boca Raton, FL, USA: CRC Press; 2021. pp. 875-889

[12] Leifer I, Lehr WJ, Simecek-Beatty D, Bradley E, Clark R, Dennison P, et al. State of the art satellite and airborne marine oil spill remote sensing: Application to the BP Deepwater horizon oil spill. Remote Sensing of Environment. 2012;**124**:185-209

[13] Fingas M, Brown CE. A review of oil spill remote sensing. Sensors. 2017; **18**(1):91

[14] Fingas M. The challenges of remotely measuring oil slick thickness. Remote Sensing. 2018;**10**(2):319

[15] Wang M, Hu C. Extracting oil slick features from VIIRS nighttime imagery using a Gaussian filter and morphological constraints. IEEE Geoscience and Remote Sensing Letters. 2015;**12**(10):2051-2055

[16] Goodman R. Overview and future trends in oil spill remote sensing. Spill

Science & Technology Bulletin. 1994; **1**(1):11-21

[17] Fingas MF, Brown CE. Review of oil spill remote sensing. Spill Science & Technology Bulletin. 1997;**4**(4):199-208

[18] Fingas MF, Brown CE. An update on oil spill remote sensors. In: Proc. 28th Arctic and Marine Oil Spill Program (AMOP). Vol. 9. Canada: Tech. Seminar Calgary; 2005. pp. 825-860

[19] Samberg A. Advanced oil pollution detection using an airborne hyperspectral lidar technology. In: Laser Radar Technology and Applications X. Vol. 5791. Bellingham, Washington, USA: SPIE; 2005. pp. 308-317

[20] Salisbury JW, D'aria DM, Sabins FF Jr. Thermal infrared remote sensing of crude oil slicks. Remote Sensing of Environment. 1993;**45**(2):225-231

[21] Shih WC, Andrews AB. Infrared contrast of crude-oil-covered water surfaces. Optics Letters. 2008;**33**(24): 3019-3021

[22] Byfield V, Boxall SR. Thickness estimates and classification of surface oil using passive sensing at visible and near-infrared wavelengths. In: IEEE 1999 International Geoscience and Remote Sensing Symposium. IGARSS'99 (Cat. No. 99CH36293). Vol. 3. Piscataway, NJ, USA: IEEE; 1999. pp. 1475-1477

[23] Chen P, Li Y, Lan G, Liu B, Zhou H. Oil spills detection and monitoring using Airborn thermal infrared remote sensing in Dalian Xingang oil pipeline explosion. In: 2012 2nd International Conference on Remote Sensing, Environment and Transportation Engineering. Piscataway, NJ, USA: IEEE; 2012. pp. 1-4

[24] Ying L, Guo-xin L, Ji-jun L, Long M. Potential analysis of maritime oil spill monitoring based on MODIS thermal infrared data. In: 2009 IEEE International Geoscience and Remote Sensing Symposium. Vol. 3. Piscataway, NJ, USA: IEEE; 2009. pp. III-373

[25] Grimaldi CS, Coviello I, Lacava T, Pergola N, Tramutoli V. Near real time oil spill detection and monitoring using satellite optical data. In: 2009 IEEE International Geoscience and Remote Sensing Symposium. Vol. 4. Piscataway, NJ, USA: IEEE; 2009. pp. IV-709

[26] Huang H, Wang C, Zhang D, Zhan S, Hong S, Wang X, et al. Combining adaptive thresholding and region filling for xylene spills detection from ultraviolet images. In: 2018 OCEANS-MTS/IEEE Kobe Techno-Oceans (OTO). Piscataway, NJ, USA: IEEE; 2018. pp. 1-6

[27] Desbiens L, Roy V, Gravel JF, Taillon Y. 2.5 W, narrow linewidth, 259.0 nm, ruggedized DUV fiber laser source for remote benzene detection. In 2019 Conference on Lasers and Electro-Optics Europe & European Quantum Electronics Conference (CLEO/Europe-EQEC). Piscataway, NJ, USA: IEEE; 2019. pp. 1-1

[28] Yin D, Huang X, Qian W, Huang X, Li Y, Feng Q. Airborne validation of a new-style ultraviolet push-broom camera for ocean oil spills pollution surveillance. In: Remote Sensing of the Ocean, Sea Ice, and Large Water Regions 2010. Vol. 7825. Bellingham, Washington, USA: SPIE; 2010. pp. 159-169

[29] Skou N, Toselli F, Wadsworth A. Passive radiometry and other remote sensing data interpretation for oil slick thickness assessment, in a experimental case. ESA SP (Print). 1983;**188**:211-216

[30] Skou N. Microwave radiometry for oil pollution monitoring, measurements, and

systems. IEEE Transactions on Geoscience and Remote Sensing. 1986;**3**:360-367

[31] Laaperi A. Microprocessor controlled microwave radiometer system for measuring the thickness of an oil slick. In: 1982 12th European Microwave Conference. Piscataway, NJ, USA: IEEE; 1982. pp. 89-93

[32] Molkov AA, Kapustin I, Ermoshkin AV, Ermakov SA. Remote sensing methods for measuring the thickness of oil/oil product films on the sea surface. Sovremennye problemy distantsionnogo zondirovaniya Zemli iz kosmosa. 2020;**17**(3):9-27

[33] Hollinger JP, Mennella RA. Oil spills: Measurements of their distributions and volumes by multifrequency microwave radiometry. Science. 1973;**181**(4094): 54-56

[34] Brown CE, Fingas MF, Hawkins R. Synthetic aperture radar sensors: Viable for marine oil spill response? In: Arctic and Marine OILSPILL Program Technical Seminar 2003. Vol. 1. Ottawa, Ontario, Canada: Environment Canada; 1999. pp. 299-310

[35] Klemas V. Tracking and monitoring oil slicks using remote sensing. In 2012 IEEE/ OES Baltic International Symposium (BALTIC). Piscataway, NJ, USA: IEEE; 2012. pp. 1-7

[36] Yang CS, Kim YS, Ouchi K, Na JH. Comparison with L-, C-, and X-band real SAR images and simulation SAR images of spilled oil on sea surface. In: 2009 IEEE International Geoscience and Remote Sensing Symposium. Vol. 4. Piscataway, NJ, USA: IEEE; 2009. pp. IV-673

[37] Skrunes S, Brekke C, Eltoft T. Oil spill characterization with multi-polarization C-and X-band SAR. In: 2012 IEEE International Geoscience and Remote Sensing Symposium. Piscataway, NJ, USA: IEEE; 2012. pp. 5117-5120

[38] Skrunes S, Brekke C, Eltoft T, Kudryavtsev V. Comparing near-coincident C-and X-band SAR acquisitions of marine oil spills. IEEE Transactions on Geoscience and Remote Sensing. 2014;**53**(4):1958-1975

[39] Marzialetti P, Laneve G. Oil spill monitoring on water surfaces by radar L, C and X band SAR imagery: A comparison of relevant characteristics. In: 2016 IEEE International Geoscience and Remote Sensing Symposium (IGARSS). Piscataway, NJ, USA: IEEE; 2016. pp. 7715-7717

[40] Collins MJ, Denbina M, Minchew B, Jones CE, Holt B. On the use of simulated airborne compact polarimetric SAR for characterizing oil–water mixing of the Deepwater horizon oil spill. IEEE Journal of Selected Topics in Applied Earth Observations and Remote Sensing. 2015; **8**(3):1062-1077

[41] Hensley S, Jones C, Lou Y. Prospects for operational use of airborne polarimetric SAR for disaster response and management. In: 2012 IEEE International Geoscience and Remote Sensing Symposium. Piscataway, NJ, USA: IEEE; 2012. pp. 103-106

[42] Laneve G, Luciani R. Developing a satellite optical sensor based automatic system for detecting and monitoring oil spills. In: 2015 IEEE 15th International Conference on Environment and Electrical Engineering (EEEIC). Piscataway, NJ, USA: IEEE; 2015. pp. 1653-1658

[43] Dan W, Jifeng S, Yongzhi Z, Pu Z. Application of the marine oil spill surveillance by satellite remote sensing.

In: 2009 International Conference on Environmental Science and Information Application Technology. Vol. 1. Piscataway, NJ, USA: IEEE; 2009. pp. 505-508

[44] Rocca F. Remote sensing from space for oil exploration. In: 2015 IEEE International Geoscience and Remote Sensing Symposium (IGARSS). Piscataway, NJ, USA: IEEE; 2015. pp. 2876-2879

[45] Minchew B, Jones CE, Holt B. Polarimetric analysis of backscatter from the Deepwater horizon oil spill using L-band synthetic aperture radar. IEEE Transactions on Geoscience and Remote Sensing. 2012;**50**(10):3812-3830

[46] Bayındır C, Frost JD, Barnes CF. Assessment and enhancement of Sar noncoherent change detection of sea-surface oil spills. IEEE Journal of Oceanic Engineering. 2017;**43**(1):211-220

[47] Xu L, Wong A, Clausi DA. An enhanced probabilistic posterior sampling approach for synthesizing Sar imagery with sea ice and oil spills. IEEE Geoscience and Remote Sensing Letters. 2016;**14**(2):188-192

[48] Brown CE, Fingas MF. New space-borne sensors for oil spill response. In: International Oil Spill Conference. Vol. 2001, No. 2. Washington, DC, USA: American Petroleum Institute; 2001. pp. 911-916

[49] Lecomte, E. En Fevrier 2017, des Drones vont Traquer la Pollution Maritime. 2017. Available online: https://www.sciencesetavenir.fr/high-tech/drones/en-fevrier-2017-des-drones-vont-traquer-lapollution-maritime_109732 [Accessed: January 1, 2017]

[50] Kirkos G, Zodiatis G, Loizides L, Ioannou M. Oil pollution in the waters of Cyprus. Oil Pollution in the Mediterranean Sea: Part. 2017;**II**:229-245

[51] Barenboim GM, Borisov VM, Golosov VN, Saveca AY. New problems and opportunities of oil spill monitoring systems. Proceedings of the International Association of Hydrological Sciences. 2015;**366**:64-74

[52] Hook S, Batley G, Holloway M, Ross A, Irving P, editors. Oil Spill Monitoring Handbook. Clayton, Victoria: Csiro Publishing; 2016

[53] Saleem A, Al Maashri A, Eldirdiry O, Ghommam J, Bourdoucen H, Al-Kamzari A, et al. Detection of oil spill pollution in seawater using drones: Simulation & lab-based experimental study. In: 2021 IEEE International IOT, Electronics and Mechatronics Conference (IEMTRONICS). Piscataway, NJ, USA: IEEE; 2021. pp. 1-5

[54] Alharam A, Almansoori E, Elmadeny W, Alnoiami H. Real time AI-based pipeline inspection using drone for oil and gas industries in Bahrain. In: 2020 International Conference on Innovation and Intelligence for Informatics, Computing and Technologies (3ICT). Piscataway, NJ, USA: IEEE; 2020. pp. 1-5

[55] Oliveira A, Pedrosa D, Santos T, Dias A, Amaral G, Martins A, et al. Design and development of a multi rotor UAV for oil spill mitigation. In: OCEANS 2019-Marseille. Piscataway, NJ, USA: IEEE; 2019. pp. 1-7

[56] Wu CH, Hsieh JW, Wang CY, Ho CH. Marine pollution detection based on deep learning and optical flow. In: 2020 International Computer Symposium (ICS). Piscataway, NJ, USA: IEEE; 2020. pp. 376-381

[57] De Kerf T, Gladines J, Sels S, Vanlanduit S. Oil spill detection using

machine learning and infrared images. Remote Sensing. 2020;**12**(24):4090

[58] Jiang Z, Zhang J, Ma Y, Mao X. Hyperspectral remote sensing detection of marine oil spills using an adaptive long-term moment estimation optimizer. Remote Sensing. 2021;**14**(1):157

[59] Yin H, Chen S, Huang R, Chang H, Liu J, Qi W, et al. Real-time thickness measurement of marine oil spill by fiber optic surface Plasmon resonance sensors. Frontiers in marine. Riga, LV: Science. 2022;**1937**

[60] Li Y, Yu Q, Xie M, Zhang Z, Ma Z, Cao K. Identifying oil spill types based on remotely sensed reflectance spectra and multiple machine learning algorithms. IEEE Journal of Selected Topics in Applied Earth Observations and Remote Sensing. 2021;**14**:9071-9078

[61] Dala A, Arslan T. In situ microwave sensors and switching circuit for oil slick thickness measurement. IEEE Sensors Journal. 2022;**22**(9):9027-9034

[62] De Carolis G, Adamo M, Pasquariello G. On the estimation of thickness of marine oil slicks from sun-glittered, near-infrared MERIS and MODIS imagery: The Lebanon oil spill case study. IEEE Transactions on Geoscience and Remote Sensing. 2013;**52**(1):559-573

[63] Jorgensen SE, editor. Encyclopedia of Environmental Management, Four Volume Set. Boca Raton, FL, USA: CRC Press; 2012

[64] Yadav M, Sharma D, Sharma O. Propagation of electromagnetic waves in multilayer structure. IJRREST: International Journal of Research Review in Engineering Science and Technology. 2013;**2**:81-82

[65] Hammoud B, Mazeh F, Jomaa K, Ayad H, Ndadijimana F, Faour G, et al. Multi-frequency approach for oil spill remote sensing detection. In: 2017 International Conference on High Performance Computing & Simulation (HPCS). Piscataway, NJ, USA: IEEE; 2017. pp. 295-299

[66] Ulaby FT, Long DG, Blackwell WJ, Elachi C, Fung AK, Ruf C, et al. Microwave Radar and Radiometric Remote Sensing. Ann Arbor, MI, USA: University of Michigan Press; 2014

[67] Hammoud B, Daou G, Wehn N. Multidimensional minimum Euclidean distance approach using radar Reflectivities for oil slick thickness estimation. Sensors. 2022;**22**(4):1431

[68] Muntini MS, Pramono YH, Minarto E, Kalsum U, Rachmanita RE. Modeling and simulation of microwave propagation on crude oil heating. In: 2017 International Seminar on Sensors, Instrumentation, Measurement and Metrology (ISSIMM). Piscataway, NJ, USA: IEEE; 2017. pp. 46-50

[69] Hammoud B, Ndagijimana F, Faour G, Ayad H, Jomaah J. Bayesian statistics of wide-band radar reflections for oil spill detection on rough ocean surface. Journal of Marine Science and Engineering. 2019;**7**(1):12

[70] Hammoud B, Mazeh F, Jomaa K, Ayad H, Ndagijimana F, Faour G, et al. Dual-frequency oil spill detection algorithm. In: 2017 Computing and Electromagnetics International Workshop (CEM). Piscataway, NJ, USA: IEEE; 2017. pp. 27-28

[71] Hammoud B, Faour G, Ayad H, Ndagijimana F, Jomaah J. Performance analysis of detector algorithms using drone-based radar systems for oil spill detection. Multidisciplinary Digital

Publishing Institute Proceedings. 2018; **2**(7):370

[72] Hammoud B, Wehn N. A maximum A-posteriori probabilistic approach using UAV- nadir-looking wide-band radar for remote sensing oil-spill detection. In: 2022 International Geoscience and Remote Sensing Symposium (IGARSS). Piscataway, NJ, USA: IEEE; 2022. pp. 7823-7826

[73] Hammoud B, Ayad H, Fadlallah M, Jomaah J, Ndagijimana F, Faour G. Oil thickness estimation using single-and dual-frequency maximum-likelihood approach. In: 2018 International Conference on High Performance Computing & Simulation (HPCS). Piscataway, NJ, USA: IEEE; 2018. pp. 65-68

[74] Daou G, Maroun CB, Hammoud B. Advanced iterative multi-frequency algorithm used by radar remote-sensing Systems for oil-Spill Thickness Estimation. In: 2021 International Conference on Electrical, Computer and Energy Technologies (ICECET). Piscataway, NJ, USA: IEEE; 2021. pp. 1-6

[75] Maroun CB, Daou G, Hammoud B, Hammoud B. Machine learning using support vector regression in radar remote sensing for oil-spill thickness estimation. In: 2021 18th European Radar Conference (EuRAD). Piscataway, NJ, USA: IEEE; 2022. pp. 221-224

[76] Hammoud B, Maroun CB, Ney J, Wehn N. Artificial neural networks-based radar remote sensing to estimate geographical information during oil-spills. In: 2022 European Signal Processing Conference (EUSIPCO). Piscataway, NJ, USA: IEEE; 2022

[77] Hammoud B, Jomaa K, Ndagijimana F, Faour G, Ayad H, Jomaah J. Experimental validation of algorithms used by radar remote-sensing systems for oil-spill detection and thickness estimation. In: 2019 16th European Radar Conference (EuRAD). Piscataway, NJ, USA: IEEE; 2019. pp. 205-208

[78] Hammoud B, Ndagijimana F, Faour G, Ayad H, Fadlallah M, Jomaah J. Reflection-coefficient experimental extraction from S21-parameter for radar oil-spill detection application. Satellite. Oceanography and Meteorology. 2018;**3**(2):5-6

[79] Hou Y, Li Y, Liu B, Liu Y, Wang T. Design and implementation of a coastal-mounted sensor for oil film detection on seawater. Sensors. 2017;**18**(1):70

[80] Texeira CC, Siqueira CY, Neto FR, Miranda FP, Cerqueira JR, Vasconcelos AO, et al. Source identification of sea surface oil with geochemical data in Cantarell, Mexico. Microchemical Journal. 2014;**117**:202-213

[81] Salberg AB, Larsen SØ. Classification of ocean surface slicks in simulated hybrid-polarimetric SAR data. IEEE Transactions on Geoscience and Remote Sensing. 2018;**56**(12):7062-7073

Chapter 4

Combining Predicted Seabird Movements and Oil Drift Using Lagrangian Agent-Based Model Solutions

Mads Nistrup Madsen, Henrik Skov and Michael Potthoff

Abstract

In traditional oil spill risk assessments, the mortality of seabirds is typically assessed based on a simulated amount of oil combined with a statistical and static (seasonal mean) number of birds within a given grid cell. The size of the cell is typically in the order of 10 by 10 km. Cell averaging in a coarse Eulerian grid will inevitably introduce a high degree of uncertainty with respect to real impact, and due to the patchiness in seabird distribution may result in over-estimation of impacts outside high-density areas and underestimation within high-density patches. Lagrangian agent-based modelling of species movements and oil drift directly would provide consistent results independent of the grid resolution and, at the same time, provide a fine-scale resolution of potential impacts. The robustness of this approach is demonstrated for a potential oil spill in the Barents Sea in an area with a high density of Common Guillemot, followed by a discussion on how this approach can improve future risk assessments during oil spills.

Keywords: oil spill risk assessments, common guillemot, Barents Sea, agent-based modelling, Langrangian modelling

1. Introduction

In traditional risk assessments, the mortality of seabirds is typically assessed based on a simulated amount of oil combined with a statistical and static (seasonal mean) number of birds within a given grid cell; the size of the cell is typically in the order of 10 by 10 km [1]. It is obvious that cell averaging in a coarse Eulerian grid introduces a high degree of uncertainty with respect to real impact, and due to the patchiness in seabird distribution may result in over-estimation of impacts outside high-density areas and underestimation within high-density patches.

As an alternative to the Eulerian approach risk assessments may be undertaken using the concept of Lagrangian particle tracking. The Lagrangian approach potentially improves the accuracy of risk assessments as the movement of oil particles can be simulated in parallel with a simulation of density and movement of seabirds using

 IntechOpen

the same weather and oceanographical model scenarios. In addition, the predicted impact will have high spatial precision as both oil and seabird particles will be resolved independently from any model grid. With the Langrangian model approach the industry standard risk assessments of oil spills [2, 3] could be further improved. Additionally, the approach would make the results of oil risk assessments in different geographical areas more comparable in the future.

The applicability of this approach in future oil risk assessments is demonstrated for a potential oil spill in the Barents Sea based on the results of the Marine Animal Ranging Assessment Model Barents Sea (MARAMBS) project.

The MARAMBS project (2018–2019) was funded by the Research Council of Norway and aimed for improving the knowledge of the distributional dynamics of seabirds in the Barents Sea. The tentative oil spill scenario was undertaken for an area with a high density of Common Guillemot (*Uria aalge*) during 30 days of the post-breeding period in September 2016.

2. Lagrangian modelling concept

Both the oil spill and the bird densities are simulated using a Lagrangian particle modelling approach. Each particle represents generally a set of variables and attached computations. This means that instead of simulating the distribution/concentrations per grid element (Eulerian model space), the density/distribution emerges as the result of particles moving across the model domain (Lagrangian space). Note that particles can move freely within the model space, i.e. they are described by a continuous point location instead of stepwise grid coordinates. Parameters like flow speeds, water depth etc. are interpolated from the computational mesh elements to the point coordinates of a particle. The differentiation between a Lagrangian oil spill and an agent based model (ABM) for the seabirds is more conceptual as both use the very same underlying techniques; an ABM model usually describes particles/agents that control their motion based on some decision-based rules computations whereas the movement in simple Larangian models is mostly passive, i.e. current driven.

2.1 Agent based modelling (ABM)

As described above the movement of an oil spill particle model mainly governed by the underlying hydrodynamics, wind forcing and weathering process. The seabird movement is the outcome of a complex agent-based model (ABM) describing the movement, feeding and internal states based on a set of input factors (forcing) and sub-models as outlined in **Figure 1** and **Table 1**. The model template used is the CBIRD ABM module in DHI's MIKE ABM Lab framework [4].

One of the main challenges related to agent-based modelling of the behaviour of individual sea-birds is to strike a balance between realistic parameterisation and heuristic representativeness of the model.

The CBIRD ABM model is described in detail in [5]. The description is repeated in brief below. It largely follows the ODD protocol for describing individual- and agent-based models [6, 7]. The ODD protocol consists of seven elements: the first three elements (purpose, entities, overview) provide an overview, the fourth element (design concepts) explains general concepts underlying the model's design, and the remaining three elements (calibration, parameterisation, validation) provide details.

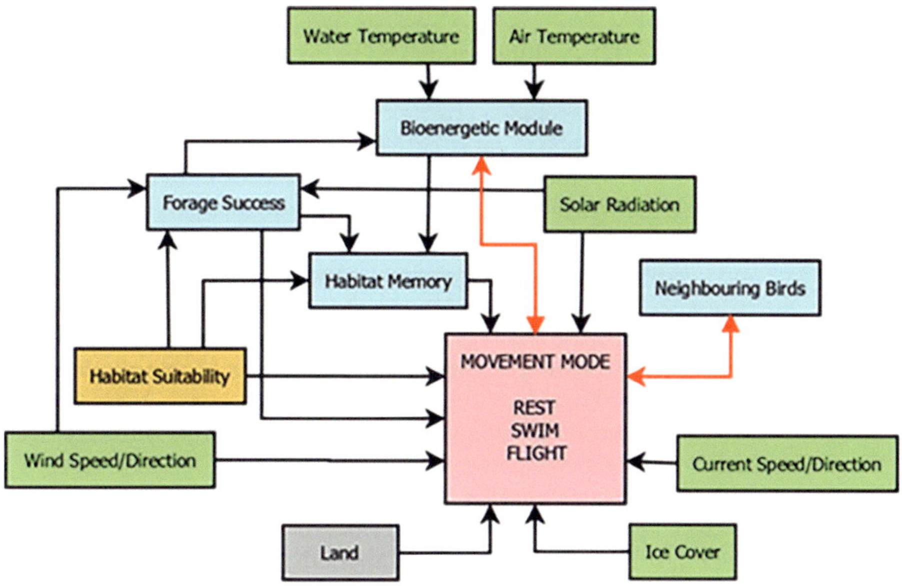

Figure 1.
Conceptual diagram of the various parameters/sub-modules affecting movement mode decisions in the envisioned seabird ABM. Green box variables denote Eulerian spatiotemporal model forcings, while blue boxes indicate Lagrangian variables/processes. Red arrows indicate two-way feedback mechanisms.

Overview of CBIRD modules with brief descriptions	
Module	**Description**
Core movements	Once sea-birds leave their colonies after the breeding season, they will switch between 3 states: resting on the water surface, swimming on or below the water surface, and flying in the air.
Bioenergetic Module	Governs physiological processes such as thermoregulation, energy expenditure of different core movements and digestion.
Day/night	Sea-birds can be set to observe diurnal behavioural cycles.
Habitat memory	Location of past encounters with preferable habitat can be readily accessed by the sea-bird. This increases overall foraging success.
Ice/land	Ice and land both represent obstacles that a sea-bird must avoid if it is to remain in comfortable environmental conditions. This module aids navigation around land masses and encroaching ice.
Wind	Unobstructed on the open ocean, sea-bird movements are directly affected by wind drift when sea-birds are on the water's surface or flying. Wind speeds and direction also affect take-off probabilities.
Social	Sea-birds exhibit flocking behaviour, and tend to move in groups to feed, avoid predation, or react to local environmental conditions.
Storm	Stormy weather can threaten sea-bird survival, especially with no shelter on the open water. Sea-birds will often move away from this perceived threat.
Moult	Some species cannot fly while moulting at sea. Besides limiting their excursion range, the start of moulting is also critical to ensure that sea-birds are not initially stranded in poor habitat patches.
Moulting area	Once leaving their colonies for moulting areas, some species are observed to be extremely targeted in their trajectory towards these moulting areas. They will fly over good habitat patches and only begin moulting in these specific areas, mainly in the Central Barents region.

Table 1.
Overview of CBIRD modules.

The purpose of the CBIRD model is to predict the dynamic spatiotemporal distribution of seabirds like the Common Guillemot in the Barents Sea by combining several individual movement and feeding behaviours as a function of explicitly modelled bioenergetics with the included effect of physical environmental forcings, such as ocean currents, wind drag, habitat suitability and ice cover.

The model tracks the horizontal position and internal state of individual seabird agents (the Lagrangian entity) inside the spatially explicit model domain. Each agent represents multiple individuals and can thus be considered a 'super-individual' [8]. The memory of previously visited habitat locations (x, y coordinates) is stored as attributes of the entity, while the strength of the habitat memory is described by a reference memory (dimensionless) to the previously visited habitat versus a satiation memory (dimensionless) relating to the perceived habitat utility (dimensionless) of staying in the current habitat. The time attribute, time since the last food encounter (minutes), controls the magnitude of swimming activity of migrating birds relative to drag forces imposed on agents by wind and currents.

All model calculations of state variables and updates of environmental forcings occur at a discrete time step over the simulation period. At the beginning of each model time step, the following sequential order is applied:

1. Update of environmental input forcings. For environmental forcings of different spatiotemporal resolution than the ABM model, spatiotemporal interpolation routines are applied

2. Calculation of process equations in the Eulerian Framework

3. Update of state-variables in the Eulerian Framework

4. Update of sensing functions for each agent relative to both Eulerian and Lagrangian frameworks

5. Calculation of Lagrangian arithmetic expressions based on updated values obtained from (1), (3), and (4)

6. Update of Lagrangian state-variables based on calculations listed in (5)

2.2 Oil spill modelling

The oil model is a multi-component model, i.e. the total oil mass is distributed among different hydrocarbon components, typically defined by their density. The masses and the properties of each particle may change over time due to weathering. Each oil component has its own weathering process. The fate of the spilled oil is typically divided into different processes:

- Drifting - (the motion of the oil caused by the ambient winds, currents and waves)

- Spreading (**Figure 2**) the motion of oil induced by its buoyancy and surface tension properties relative to water. Oil spilled on the water surface immediately spreads over a slick of few millimetres in thickness. The spreading is primarily promoted by gravity and surface tension; however, many spills of varying sizes quickly reach a similar average thickness of about 0.1 mm.

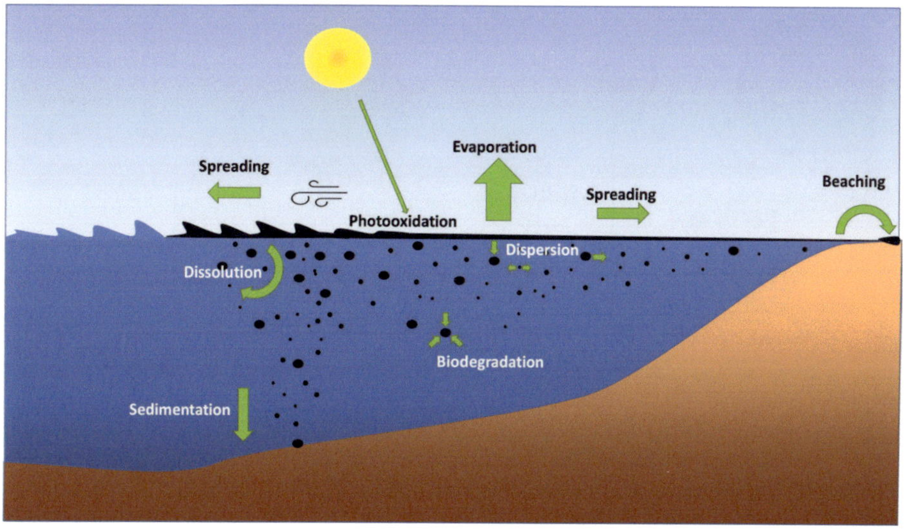

Figure 2.
Processes acting on spilled oil, (modified after [9]).

- Weathering (**Figure 2**) processes causing physical and bio-chemical changes of the oil by evaporation, emulsification, biodegradation, dispersion, dissolution, oxidation, and sedimentation and beaching. Spreading, evaporation, dispersion, and dissolution can be defined as short-term weathering processes, whereas emulsification, biodegradation, and photochemical oxidation are recognised as long-term weathering processes (**Figure 2**).

2.3 Interaction between agents

A risk screening is carried out for Common Guillemot based on the oil spill model results by computing the spatial overlap of the oil slick and the simulated species distribution. Each oil particle represents a given area with a given slick thickness and relevant properties such as oil viscosity and depth below the water surface. The latter determines whether the oil floats on the surface or the oil droplets are dispersed into the water column due to wave action.

For each time step in the oil spill simulation period, it is tested if the position of the individual sea bird agent particle is within the oil area represented by the individual oil spill particles. If this is true, the sea bird agent particle is flagged as being in contact with oil. However, it is further tested before flagging if the following parameters exceed threshold levels for the sea-bird in question.

- Diving depth (Zmin). If the oil particle is below Zmin, there is no flagging.

- Oil slick thickness. If the oil slick thickness is lower than the threshold level, there is no flagging.

- Viscosity (cP). If the viscosity is higher than the threshold value, there is no flagging.

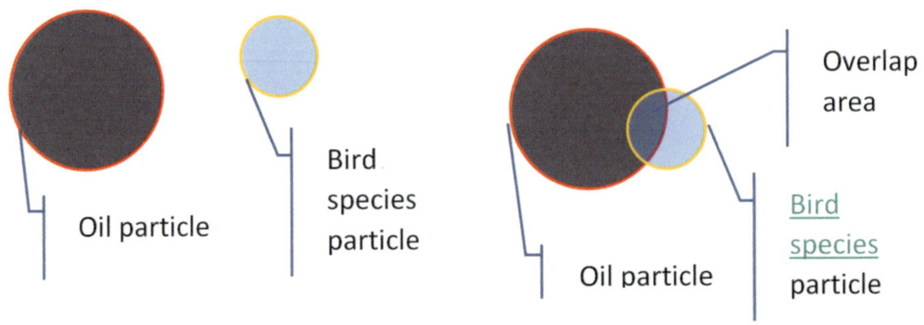

Figure 3.
Schematic of the calculation of the affected number of individuals based on the overlap area of the oil- and bird species particle.

Each sea bird particle/agent represents a large number of individuals, i.e. it is a so-called super individual. This must be considered when assessing whether or not individuals of a species have been in contact with a drifting oil slick. In case of super individuals, the point position of the agent/particle is replaced with an area around the particle. It is assumed that the individual birds represented by the super individual particle are uniformly distributed within this area. If there is a spatial overlap with the area defined by any oil particle, the affected number of individuals corresponds to the relative overlap area of the bird species particle. The area covered by the bird species particle is then reduced proportionally or, when no unaffected individuals remain, the particle is removed from the analysis. This means that the number of affected birds serves as a worst-case estimate (**Figure 3**).

The risk screening for the oil spill is then expressed in terms of the number of birds exposed (flagged) to oil above the threshold values during the oil spill simulation period. The number of birds exposed is also expressed as a fraction of the average population size with the model domain during the oil spill simulation period. The oil spill simulation will typically continue 30 days after the spill has terminated.

The above approach is based on pure Lagrangian ABM results for birds and oil, respectively, which in turn provide consistent results independent of the Eulerian grid mapping (i.e., results that are independent of the grid resolution).

3. Case study in the Barents Sea

The hydrodynamic data and most forcing parameters for the ABM and Oil spill model are driven by DHI's existing 3-dimensional hydrodynamic model. The model domain, encompassing the Barents Sea and the colonies in Norway, Russia and the islands of Svalbard, Franz Josef and Novaya Zemlya is shown in **Figure 4**. The modelling software package applied was DHI's 3-dimensional flexible mesh model, MIKE 3 FM [10] which includes both meteorological, tidal and oceanographic effects.

3.1 Common guillemot

Common Guillemot is a large auk species with a circumpolar distribution, breeding in dense colonies between 40°N and 75°N. Like other auk species, the wings are used for both swimming and diving. Common Guillemots are excellent divers with

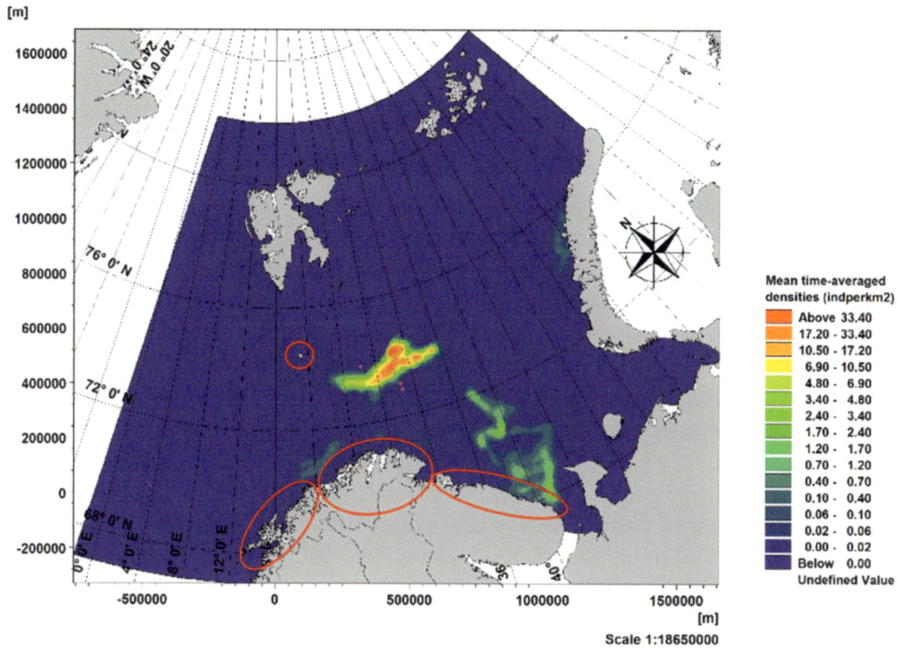

Figure 4.
Common guillemot modelled mean density September 2013. Pink dots indicate common guillemot survey sightings. The red circles/ellipses encompass the main colonies. Model domain extent in stereographic projection (Stereo_1670_Sc1).

maximum diving depths of more than 200 m. It is a pelagic pursuit diver, and the primary food item in the Barents Sea is small pelagic fish such as capelin and fish larvae and fry. Common Guillemots leave the colony in late July - early August. The male follows the chick for the first two months at sea. During this period, they moult and are flightless for approximately 45–50 days.

The population of Common Guillemots in the Barents Sea has declined dramatically since the first censuses in the 1960s. In particular, the mass mortality of birds during the winter of 1986–1987 decimated the populations along the Norwegian coast and on Bjørnøya. This incident was probably caused by food limitation due to a crash in the capelin stock and very low densities of alternative food items. Since then, the populations have been growing, but the species is still listed as critically endangered on the Norwegian Red List. Colonies are found in the southern part of the Barents Sea. The Barents Sea populations are year-round residents to the southern Barents Sea area, and in addition, Common Guillemots from Norwegian colonies further south migrate into the Barents Sea during autumn.

The population of Common Guillemot of 300,000 pairs is concentrated in colonies around Bjornoya/Bear Island (83.9% of the population) and the Finnmark Coast (8.9%), with minor colonies at the Murman Coast (5.4%) and in the Norwegian Sea (1.8%). *https://www.npolar.no/en/species/common-guillemot/

3.2 Oil spill

For illustration, the oil from a hypothetical oil spill in the Barenst Sea was applied as a representative oil type with a spill rate corresponding to 700 m^3/day and a spill duration of 8 days. The spill is assumed to be a topside surface spill.

The spill was released on 1st September 2016 at (73° 34′ 40″ N, 22° 55′ 5″ E), and the simulation continued 30 days after the spill had terminated.

3.3 Mortality assessment

The sensitivity of different sea-bird species to oil pollution is described as a function of the relative amounts of time spent on the sea surface, the importance of the population within the study area and the size of the world population. The knowledge underpinning these parameters is relatively well established due to large amounts of standardised data from countrywide surveys of beached dead sea-birds, surveys of sea-bird densities at sea, counts of the number of breeding sea-birds in colonies and information on specific details of breeding and survival.

Thresholds of relevant species of sea-birds in the Barents Sea to the characteristics of oil residues (key effect features) were provided. The key effect features of oil in relation to sea-birds are film thickness, viscosity, and depth of the oil in the water column (**Table 2**). Although the effect in relation to depth in the water column can be confidently assessed from general knowledge of the diving capacity of the different species, the actual diving depth per species may fluctuate depending on season and location. It may differ in the Barents Sea from other areas. More importantly, the thresholds related to film thickness and viscosity have been established without undertaking tests on live birds [11]. Hence, they should be regarded as guidelines rather than well-defined thresholds.

The modelled distribution of Common Guillemots showed peak occurrence at the boundary between the Norwegian Coastal Current and the Atlantic water mass. The predicted (mean) distribution of suitable Common Guillemot habitat displays a well-defined concentration 150–200 km north of the Norwegian coast. Both oil and Common Guillemots displayed a high degree of spatio-temporal variation which resulted in a limited intersection between the modelled oil slick and the main concentration of the guillemots (**Figure 5**). The intersection took place in the area just south of Bjørnøya. An animation of the 38 day simulation period can be displayed here: Video: https://www.youtube.com/watch?v=lgzBrfiQm6g.

The estimated hourly number of impacted Common Guillemots is displayed in **Figure 6**. Due to the limited intersection between the oil and guillemot particles the guillemots were only impacted during three of the 40 days in the model period (2–4 September 2016). Further, the modelled impact only took place during 15 of the 213

Species	Zmin (m)	Oil slick thickness	Viscosity (cP)
Glaucous Gull	<20	>10 μm	<3000
Black-legged Kittiwake	<20	>10 μm	<3000
Common Guillemot	<100	>10 μm	<3000
Brünnich's Guillemot	<100	>10 μm	<3000
Atlantic Puffin	<50	>10 μm	<3000
Little Auk	<30	>10 μm	<3000

Table 2.
Example of effect features of oil residues relevant for assessing impacts of oil on sea-birds in the Barents Sea. Zmin is the depth at which the species will be at risk from oil.

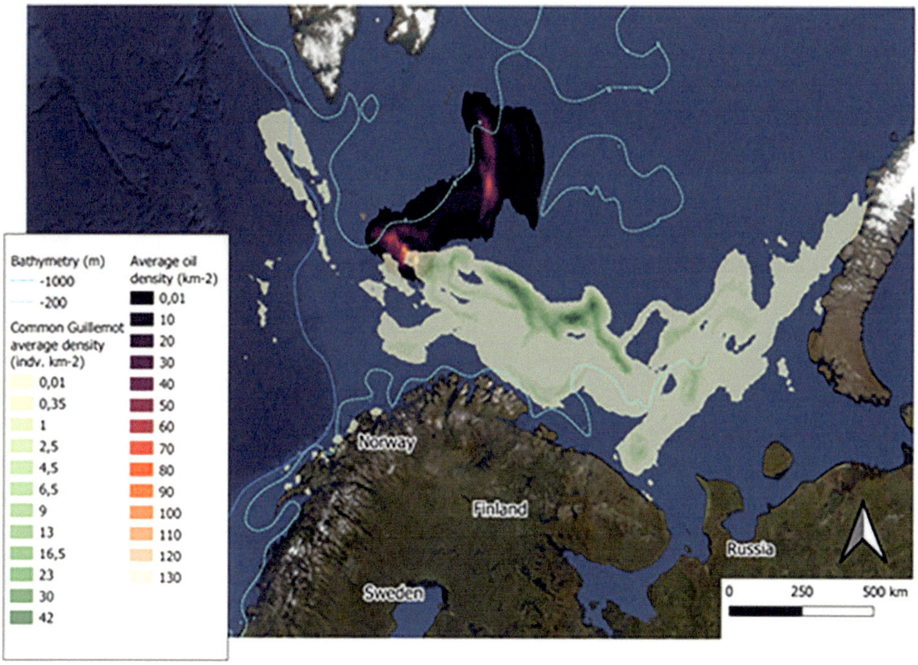

Figure 5.
Overlap between modelled mean oil density (L/km²) and modelled mean density (n/km²) of common guillemot during the test period. - "sources: Esri, DigitalGlobe, GeoEye, i-cubed, USDA FSA, USGS, AEX, Getmapping, Aerogrid, IGN, IGP, swisstopo, and the GIS user community".

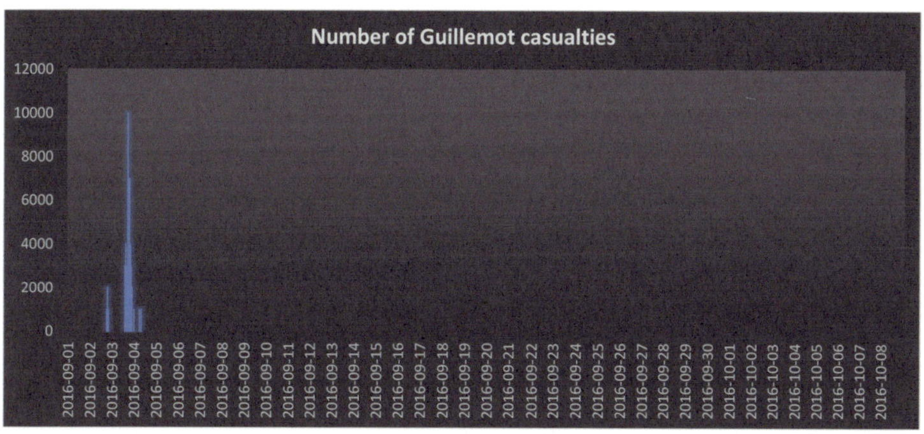

Figure 6.
Estimated hourly number of casualties of individual common guillemot during the oil spill simulation 1/9–8/102016.

modelled hours equivalent to 7% of the trajectory time. The total number of impacted guillemots during the modelling period was 50,000. With a total number of breeding pairs of 300,000 the number of Common Guillemots in the western Barents Sea during autumn 2016 would be approximately 1 million, taking juveniles and non-breeding immatures and adults into account. Thus, the relative impact on the population represents 5% of the population (**Figure 7**).

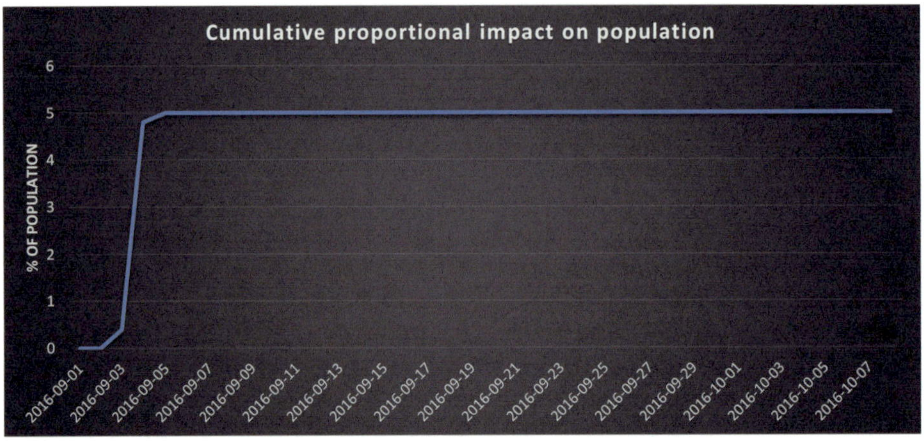

Figure 7.
Cumulative impact of simulated oil spill on common guillemot population in the Barents Sea 1/9–8/102016.

4. Conclusion/discussion

This demonstration has stressed the potential for applying a combined fine-scale Langrangian modelling design for improved prediction of the movements and intersections between an oil spill and seabirds. The oil spill risk modelling method provides a powerful tool for an initial screening of the potential impact on the various seabird species during a given period, even if the approach does not comply with a full risk assessment in line with the traditional MIRA class approach in Norwegian Waters [1]. Furthermore, combining the ABM model results of species movement directly with the result of oil particle movements provides consistent results independent of the grid mapping (i.e. results that are independent of the grid resolution). This makes results from different areas directly comparable.

More importantly, studies of the marine distribution of birds unambiguously point at the fine-scale distribution of most species of seabirds. Seabirds predominantly show an affinity to physical oceanographic properties such as fronts, upwellings and eddies, which enhance the probability of predators encountering prey [12–14]. In the Barents Sea this tendency is reflected in the ubiquitous concentrations of seabirds in the region of the Polar Front [15].

To accurately describe the overlap between the distribution of seabirds and oil slicks over time, one needs to be able to take account of the high degree of clustering and habitat association seen in seabirds. This solution can only be achieved with a Langrangian modelling approach like the one tested here. By using a standard Heuristic approach based on mean seasonal distrubutions of seabirds predicted impacts will unlikely resolve the true intersection in the distribution of the oil slick and the seabirds.

In other words, if high-resolution Langrangian models are not applied as part of the risk assessment of oil incidents mean values rather than in situ values for oil and seabirds predicted intersections will rarely match reality. As a result, risk assessments may lead to a type II error—a result estimating an impact in an area of low seabird density —or a type I error—a result erroneously pointing at a smaller or medium impact in an area where seabirds are highly concentrated. Thus, despite the large number of risk assessments of oil spills undertaken in the past accurate assessment of the impacts of oil slicks on seabirds remains a challenge.

The combined fine-scale Langrangian modelling design seems to have a strong potential to pave the way for more realistic assessments of the concurrent distribution and movement of oil slicks and sensitive species of seabirds. Obviously, the calibration of the both the oil spill and seabird models requires that detailed empirical data are available regarding i) the oceanographic properties of the spill site, ii) the chemical composition of the oil and iii) the local density and distribution of the target species of seabird. Spatially refined assessments of the risk of seabird species sensitive to oil pollutions will enhance both the planning and environmental management of oil and gas exploration activities in the future.

Acknowledgements

Special thanks to Equinor, ConocoPhillips and TotalEnergies for funding and valuable discussions during the development of the MARAMBS data portal and special thanks to the Norwegian Institute for Nature Research (NINA) for participation in the development of the data portal in relation to seabirds.

Author details

Mads Nistrup Madsen, Henrik Skov and Michael Potthoff*
DHI A/S, Hoersholm, Denmark

*Address all correspondence to: mpo@dhigroup.com

References

[1] OLF. Metode for Miljørettet Risikoanalyse (MIRA). Høvik, Norway: DNV Energy; 2007

[2] OILMAP, RPS-ASA. Available from: http://asascience.com/software/oilmap/

[3] OSCAR, SINTEF. Available from: https://www.sintef.no/en/software/oscar/

[4] MIKE ABM Lab Framework, DHI. Available from: https://www.mikepoweredbydhi.com/products/abm-lab

[5] Skov H, Theophilus TZE, Heinänen S, Fauchald P, Madsen M. Jonas Brandi Mortensen, Thomas Uhrenholdt, frank Thomsen: Real-time predictions of seabird distribution improve oil spill risk assessments. Marine Pollution Bulletin. 2021;**170**:112625. DOI: 10.1016/j.marpolbul.2021.112625

[6] Grimm V, Berger U, Bastiansen F, Eliassen S, Ginot V, Giske J, et al. A standard protocol for describing individual-based and agent-based models. Ecological Modelling. 2006;**198**(1-2):115-126

[7] Grimm V, Berger U, DeAngelis DL, Polhill JG, Giske J, Railsback SF. The ODD protocol: A review and first update. Ecological Modelling. 2010;**221**(23):2760-2768

[8] Parry HR, Evans AJ. A comparative analysis of parallel processing and super-individual methods for improving the computational performance of a large individual-based model. Ecological Modelling. 2008;**214**(2-4):141-152. DOI: 10.1016/j.ecolmodel.2008.02.002

[9] Fate of Marine Oil Spills. Technical Reports. London: The International Tanker Owners Pollution Federation Limited (ITOPF); 2002. p. 8

[10] MIKE 3 FM, DHI. Available from: https://www.mikepoweredbydhi.com/products/mike-3

[11] French-McCay D. Oil spill modeling for ecological risk and natural resource damage assessment. In: International Oil Spill Conference Proceedings, Portland, USA. 2011

[12] Schneider DC, Duffy DC. Scale-dependent variability in seabird abundance. Marine ecology progress series. Oldendorf. 1985;**25**(3):211-218

[13] Skov H, Prins E. Impact of estuarine fronts on the dispersal of piscivorous birds in the German bight. Marine Ecology Progress Series. 2001;**214**:279-287

[14] Fauchald P, Skov H, Skern-Mauritzen M, Hausner VH, Johns D, Tveraa T. Scale-dependent response diversity of seabirds to prey in the North Sea. Ecology. 2011;**92**:228-239

[15] Madsen M, Heinänen S, Mortensen JB, Mortensen LO, Theophilus TZE, Skov H, et al. MARAMBS – A Web-Based Software Tool to Assess Movements of Marine Life in the Barents Sea. Commissioned by the Research Council of Norway. Hørsholm, Denmark: DHI Report; 2019

Chapter 5

Ballast Water Utopia and Some Environmental Protection Ideas

Fikile Portia Ndlovu

Abstract

An environmental protection plan that would resemble a true utopia might still be distant however, employing advanced methods and innovation that will get the global community on the road to the best marine and other environmental protections should be every generation's intent. Ballast water management as related to shipping, trade, and the marine world as a special concern of this paper has been one of the most notable innovation drivers in the maritime space in the last couple of decades in order to close doors to the dangers of introducing new aquatic invasive species in sensitive environments around the world (bio invasions). This paper aims to selectively survey some established ballast water management methods reimagined by the latest innovations reflecting on some latest developments. The world is constantly facing troubling concerns over environmental protection issues particularly related to marine and maritime related endeavors. For example, what it may mean to lose many Russian scientists that are involved in arctic research as a result of the Russia-Ukraine conflict, especially with ship traffic and ballast water management slated to be more and more of a concern in that area as shipping channels become more open in that region.

Keywords: ballast water management, ship tech, marine pollution, green ship, solutions to invasive species, Arctic shipping lanes, scientists in Arctic research, pollution monitoring

1. Introduction

The term "Ballast Water Utopia" is not the author of this paper coined phrase but a term stated by William Burroughs of Freedom Ballast, a company that provides an innovative solution of ballast water treatment as a barge or portside service [1]. The strong language suggested by the term "Ballast Water Utopia" by the company's CEO inspired the discussion of cutting-edge innovations in ballast water management in this paper. To begin with, this paper's use of the word utopia attests to the commitment that certain innovators have in matters related to environmental protection especially in the marine world by focusing on ideas supported by technology that could potentially lead to the most ideal situations in dealing with invasive aquatic species (AIS) in shipping technologies for ballast water management. The International Maritime Organization (IMO) [2], dedicated over a decade of momentous efforts to react to the mission to control, prevent and close the door to aquatic invasive species that were or would be introduced by ships to foreign waters through ballast water tanks.

When empty, ships need to be ballasted (balanced) with water however that water needs to be thrown overboard when cargoes are taken aboard the ship and this is where invasive species and sediments (which may be teeming with organisms) from different parts of the world may find their way to a new world where they do not have natural predators and thus are able to grow into dangerous and environmentally threatening levels, at least that is how it would be if no care is taken whatsoever to treat foreign water containing invasive species [3].

The dangers of AIS from ballast water tanks are well documented [4] having the potential of destroying entire ecosystems, entire fishing industries, and various environments [5]. Some of the most infamous invasions include the zebra mussel, which costs millions of dollars in clean-up operations in the Great Lakes of the United States. This mussel is native to the Black Sea, the Caspian Sea, and waters just south of Russia and Ukraine [6]. Further examples of notorious marine bio-invasions include the green crab, the Cladoceran Water Flea, certain jellyfish, the lionfish, etc. In a post-Covid lock-downs world, with the exception of certain areas around the world that still hold on to lock-downs, humanity, in general, in 2022 is even more aware of the dangers of uncontrollable organisms working against the life spans and the quality of life of human beings. This means that there will always stand a moral obligation to respond expeditiously to environmental threats so that commercial activity does not hinder and destroy human being, their livelihoods, and the environment.

The ocean is constantly under monitoring through various scientific undertakings around the world. Scientists warn us through data-supported, intense research that the ocean faces many environmental challenges and these are studied under varied branches of study, to name a few examples, the ocean has environmental challenges with, coral bleaching [7], marine neurotoxins termed the red tides [8], noise pollution from ships [9], concerns about the devastations that can arise from the melting permafrost, particularly in the Arctic region [10] which would be devastating to communities in that environment and the globe, not to mention all the pollution from microplastics and other effluents. In this paper, we are just focusing on the ballast water management field but this must be read within the context of other environmental challenges that scientists are looking at. This is why we cannot afford to lose scientific minds in environmental threat management. Since this paper is concerned with ballast water management, it is important to consider the ballast water simulations [11] in regions that are changing as a result of global warming such as the Arctic.

When international rules for ballast water management were initially drafted, technologies to enforce such rules needed to be brought up to date with the demands of the rules. The rules were eventually adopted and came into force in the form of the International Convention for the Control and Management of Ships' Ballast Water and Sediments (the Ballast Water Management Convention, 2004), also known as the BWM Convention, 2004 [12]. On 8 September 2017, this law came into force, with approved systems finally being available in the ship technology market. These approved systems are available and listed with the IMO and several port authorities around the world [13].

In this paper, there will be a consideration of some of the newer methods such as barge and port side ballast water treatment, as part of a survey of the latest developments in ship techs for ballast water management, there will also be a selective consideration of the effectiveness of a few sample ballast water management systems, as it would be impossible to discuss them all in this paper. Further, with the Arctic opening up more as a result of global warming and better ships and legal codes are developed to traverse that environment, naturally, concerns about ballast water management also increase in that region. It will therefore be appropriate in this paper to also briefly consider what the loss of Russian scientists may mean for Arctic research and environmental protection endeavors as a result of the war between Russia and Ukraine. Further, a consideration of other bodies of water in relation to ballast water such as the Mediterranean Sea and Antarctica will also be considered as examples of how ballast water management has affected those areas.

2. The research methods and methodology

In order to avoid a superficial discussion of the concerns around the dangers of discharging untreated ballast water into the seas as a result of international trade or other maritime-related activity, it is essential to discuss the strategy for collecting data for an analysis used in this paper. This will assist the reader to determine the reliability and validity of the findings and observations in this paper. Before discussing the strategy, it is important to note that the research question to be answered in this paper may be couched as follows, 'Following the IMO treaty of Ballast Water Management, the BWM 2004 [14], which methods have actually been considered successful and with the success of those methods, which of them can be considered as new developments that solve problems around ballast water management?' Coupled with this question is, "What are some of the actual problems associated with ballast water management especially in sensitive ecosystems such as the Arctic, Antarctic and ancient traditional trading routes such as the Mediterranean Sea, as examples to sample ballast water management concerns as well as strategies?"

To answer these questions a combination of qualitative as well as quantitative methods have been employed in this research for data collection and analysis. Apart from the author's survey which was directed at various companies that deal with ballast water development tools to determine the attitude and preparedness to comply with the new compliance standards of the BWM, 2004, the data studies in this paper to answer these questions took the form of observational, experimental literature reviews as secondary data of experts in the field. For example, in the excellent research experiment conducted by Rosenhaim et al. [11] a study of movements of ships during various seasons and how it affects the accumulation of environmentally threatening discharges of ballast water is relied on to study, analyze and problem solve in the Arctic region. Similar research methods are employed in this paper.

In order to study industry attitude and preparedness for compliance with the BWM, 2004 standards and the moral sensitivity to the seriousness of ballast water threats, the author created a survey that contained seven multiple choice questions with a Likert scale response [15], with an 8th open-ended question asking maritime industry and developers or users of ballast water technologies if they had any comments about the ballast water regulations that were about to come into force as international law. Participants were given at least 5 minutes to answer the surveys anonymously. A hundred surveys were sent to various companies and a maximum of 20 surveys were completed. All returned surveys were considered to sample industry attitudes and preparedness [16]. The surveys represent those who have been following the alarming problem of untreated ballast water discharges from ships and have a direct interest in the developments and technologies to deal with the problem.

The theoretical positioning of this research paper shows the true power of collaborative data production and interdisciplinary engagement to solve environmental concerns and problem sets. Although some of the literature review is secondary in nature, it is provided by those who have strong sound paradigms and compelling experimental results. Although the limitations of relying on secondary sources of data may be a reality, secondary sound data is far better that no data at all and it forms a powerful tool in making arguments such as the necessity of monitoring the data, if it exists for successful compliance with the ballast water management regulations which are now fully enforceable international law. The process and steps for data collection and analysis in this paper are logical, justifiable, and allow the questions raised above to be answered reasonably or raise further relevant questions that will lead to better future solutions.

3. Understanding the dangers of discharging untreated ballast water

It is important to note that a safe ship has to maintain balance by having water pumped into its ballast tanks. This gives the vessel stability as the weight of the vessel and its displacement in the water have to be carefully managed throughout the voyage which is all part of vessel stability. It has been said that water-based ballast systems and ship ballasting increased with the advent of steel-hulled ships [17]. Ships with ballast water systems are not only those that engage in the carriage of goods, for example, the dredger, ILembe, a Hopper Dredger flying under the South African flag, registered under IMO number 9741891 has a ballast water system which pumps in or out ballast water depending on how much sand and other by-products of dredging are taken on board the vessel. To keep the dredger from transporting invasive species around the coastline upon which it operates, the dredger's operations in law require that it exchanges its ballast water with local water before it gets into the new coastal waters [18]. This example shows us that ballasting may affect and type and size of vessel and the introduction of foreign waters is something that must as a result be managed carefully.

In international trade, **Figure 1** below shows us that ballast water discharge into the environment as shown in the illustration, in this case, it introduces invasive species into a new environment if such ballast water being discharged is not sufficiently

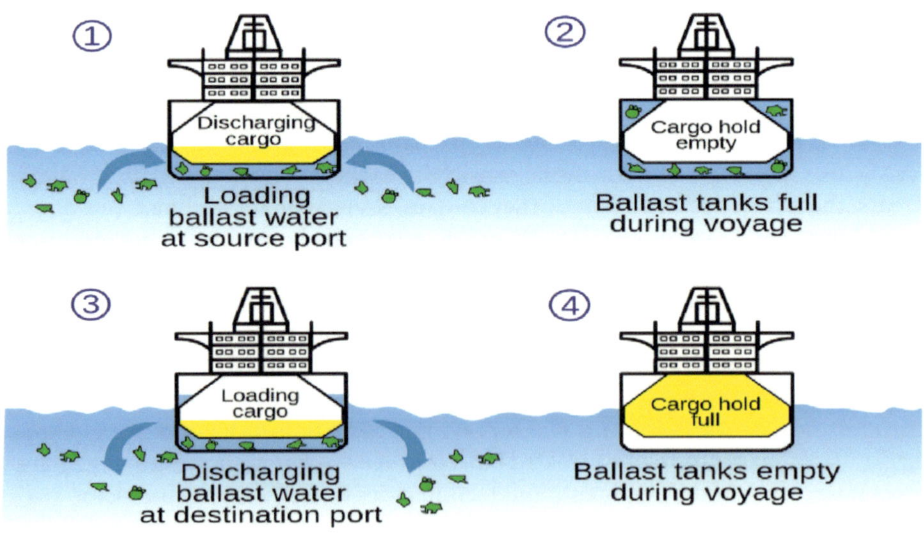

Figure 1.
Ballast water management and its implications (Miami shark research, 2016).

treated [19]. This discharge of ballast water is part of ensuring that the vessel can load cargo at the port of loading in a safe manner such that the illustration shows how the water is replaced by the cargo and vice versa. In trade, a cargo empty vessel is ballasted by water and a fully cargo-loaded vessel is ballasted by a combination of cargo and the appropriate amount of water as the operations of the vessel require. On board are sensitive ballast water systems fitted with alarms and are carefully monitored by the captain and his or her crew who give the appropriate instructions for what is considered the appropriate levels of ballasting is needed for each unique situation.

It has been established by the scientific community that ship ballasting seems to work best with water and that coming up with a technology of ballast-less ships is probably a pipe dream because the current methods are safer, economical, and have shown ship construction soundness [20]. Unfortunately, the science of avoiding invasive species from entering new worlds with shipping as a vector moved along at a slower pace when compared with just how many invasive species were historically and more recently being introduced around the world.

Untreated ballast water can introduce not only dangerous robust nonindigenous species to a new environment, but these species tend to cause economic and environmental harm by creating self-sustaining, dominant populations that disrupt the indigenous environment. Some invasive species demonstrate *allelopathy*, a condition that allows invasive species to produce chemicals that inhibit the growth of other organisms, particularly in plants, and this is of particular importance in cases of seaweed invasions. This means that billions in financial losses occur in fishing and other aquaculture sites as a result of ships with untreated ballast water bringing in parasites, viral and bacterial pathogens that may cause cholera and other cysts of toxic dinoflagellates which can be poisonous [21]. The IMO's GloBallast [22] initiative which recorded this negative phenomenon around the world also keeps records of invasions by country as a public record. This report contains alarming data which is why the BWM 2004 Convention was eventually prepared and published as is now in force. This is also the reason, the author ran a survey to find out what industry attitudes were concerning readiness to deal with untreated ballast water and to ask those within the industry if they have thought of methods not only to manage new ballast water invasions but in future hopefully deal with current invasions with an aim to reduce them significantly with the hope of reviving damaged ecosystems and other affected fishing businesses, etc.

Here are the general findings of that survey which shows that the industry leaders in this sector have not only understood the dangers of ignoring ballast water management but are keen to turn to technologies and other methods to manage ballast water AIS.

- 'The survey attracted the attention, answers, and comments from a significant portion of persons directly and heavily involved with ballast water management as their regular occupation.

- The survey also attracted members of the public who care about the marine environment, who may not be exposed to the general daily technicalities of ballast water management matters but they are concerned with potential destruction of the marine environment through shipping activities.

- A significant portion of shipowners and port officials are prepared to implement the new Convention standards (as of now, the regulations have been in force for over 3 years already).

- Decision makers in companies and ports that have to implement the standards are well aware of the new regulations and have been ready to implement them.
- In terms of financial preparedness, a smaller portion of shipowners seemed ready to spend higher capital investments on their ballast water management systems.
- A significant portion of shipowners are ready to implement the standards but do not wish to automatically spend high capital investments installing expensive ballast water treatment systems.
- A significant portion of shipowners are financially prepared for investing in compliant ballast water systems but seem suspicious of overly budget-friendly systems while some owners are most definitely interested in budget-friendly ballast water treatment systems.
- Morally participants care for the environment and wish to take responsibility for the environment being preserved. This restores hope in the technicalities of international trade.
- Unfortunately, there is a call for something to be done by the invasions that have taken over now but there is also a concern that perhaps it is too late to do anything about existing invasions.
- Many shipowners believe in the D2 standard for future compliant ships and believe in its effectiveness and a smaller portion of shipowners do not only believe in the D2 standard readiness but they are also currently prepared. It is submitted that such shipowners probably have newer ships and have ordered compliant ballast water systems for such ships in time.
- Participants who wrote some comments showed serious concerns about doing something about current invasions, some called for standardization throughout the globe so there is no weak link in ballast water management across the globe while some participants warned shipowners to be careful not to invest in a system that may not pass because it simply is not effective. Last but not least, the survey results showed serious concerns about accurate monitoring [16]. The survey results show that the BWM 2004 standard is far more prepared for by the maritime industry and this is why it is essential to discuss some of the new methods and developments in ballast water management and consider the data following the BWM 2004 regulation coming into force.

4. New methods: Barge and port side ballast water treatment

In a now-published book in 2022 written by the author of this paper with several authors who are also researchers, scholars and practitioners, in the field of ballast water management [16], it is recorded that the ballast water technologies and innovations tabulated in **Table 1** below are currently available in general on the market and they are usually employed by various shipowners. These methods have been passed by the IMO [23, 24]. The IMO has rigorous standards established in sound standards for

Physical Filtration Systems
Screens/Discs
Hydrocyclone
Coagulation
Magnetic Field Treatment
Chemical Disinfection Ballast Water Treatments
Oxidizing Biocides
Non-Oxidizing Biocides
Ultra-Violet Treatment Method
De-Oxygenation
Electric Pulse and Plasma Treatment (also available on the market is Acoustic (Cavitation Treatment) and Thermal (Heat) Solutions

Table 1.
Table of ballast water treatment systems on the market.

passing these technologies however many of them still need to be further developed and improved their effectiveness double-checked against data available after years of employment of these methods. The use of chlorine in ballast water treatments, for example, which has been one of the more simpler and popular methods of ballast water treatment, has been criticized for its potentiality to form toxins in the water because, further chlorine still carries the status of extremely dangerous chemical because in its gas form, it is so harmful that it was a chemical weapon in the first world war [25].

The IMO and ballast water management of just laws of nations around the world obviously seek only safe methods of managing ballast water to comply with IMO standards [26]. This is why it is crucial to have an understanding of technologies as recorded in **Table 1** below. This table provides us with performance data for technologies that have been popularized by the industry in its zeal to comply with the BWM 2004. The fact that there are complaints about some of the methods being considered 'dangerous' for human health means that we still have to work more on compliance methods as well as monitoring. Otherwise, this means that we are still ignoring environmental impact problems created by shipping and the necessity of trade.

In **Table 1** below the technologies which could be considered new generation to make compliance with the BWM 2004 Convention possible seem to be more popular because they deliver the D2 standards for water approval for safe discharge. These technologies also provide for easy monitoring or more developed monitoring systems which the survey conducted showed was of particular importance to the industry. The monitoring of ballast water has had its traditional systems for testing and monitoring that have been part of the shipping industry however now that the standard is raised for the management of ballast water discharges it is clear that a customer, shipowner who buys into the popular new technologies in **Table 1** will have a great chance of being able to prove ballast water management compliance.

Innovators in the ballast water management space have shown us according to the table above that technologies in the market are capable of achieving regulatory compliance for IMO and Port State control, therefore let us consider how these methods have been technically and operationally advanced to provide ballast water treatment shore side or by barge as an example of the latest developments. It is important to consider how this innovation is contributing towards a greener ship, port and ballast water management system. Green shipping is the future because without a stable environment there is no life, commerce, or trade to sustain. This cannot be ignored because it leads to devastation. We must applaud, support, and encourage those who use modern technologies for greener safer shipping [27].

The leading innovators on barge and port-side ballast water management are the company, Freedom Ballast [28]. The technology used is illustrated in the company's patented system which works as follows before ballast water is discharged from a ship it passed through the patented and patent pending UV and pasteurization technology of Freedom Ballast. This technology without the use of active substances and chemical compounds treats the ballast water before it is discharged. The system itself works with high throughput at 1000 M_3/hour, this would please the shipowners seeking that just-in-time speed for port operations.

Upon the author of this paper is granted an interview with the CEO of Freedom Ballast, it was clear that the CEO was focused on technologies to stop invasions before the technology to deal with current and existing invasions is discussed. In his discussion, the CEO discussed a world where ship hulls can be repurposed to provide logistical possibilities for this system to be available at ports throughout the globe. This means that there is much research and development behind this company to make these green ideas of not toxic, effective, recycling ballast water management to reach their full potential. The Freedom Ballast system is accepted in North America with concessions to operate in the lower Mississippi. What will be interesting to see is whether or not this technology is suitable for the mega-size class vessels and if it were to be used in ultra-sensitive marine environments like the Arctic region whether or not it would be one of the most acceptable methods. Considering the non-toxic nature of the technology, this system might be one of the latest applications that may be most suitable.

It is submitted that the spirit of environmental protection and innovation in marine environments should continue as shown with the barge port side ballast water treatment system demonstrated above. In US law, which we may use as a significant example of a global benchmark for regulatory compliance on ballast water management, new innovations in ballast water management, whether local or foreign are supported. In the case of foreign innovations and other unique methods, a request to pass these under the United States Coast Guard (USCG) is encouraged. This mean that practically it is required that a formal request is made to the USCG to decide on any new Ballast Water Management Systems (BWMS) approved by a foreign administration. This is why it can be said that standards for BWM in the US are high, see the Code of Federal Regulation particularly, Title 33 CFR, '§ 151.2026. It is submitted that governments around the world should as research grows and data becomes available in the area of ballast water management that an innovatively open mind to even better technologies be kept to continue to encourage the finding and investment in the best technologies.

In the US, again, for example, to illustrate this innovation encouraging the best solution for ballast water management strategies, when a shipowner wants to use or is using an Alternate (ballast) management systems (AMS), as long as such a system is in agreement with the aims of the BWM 2004 Convention and request to use it in writing is made to the USCG, that system can be approved and usable in the USA subject to the discretion of the port authority, of course. This means constant research and development is being encouraged in the USA, this is the commitment to the environment. It is submitted that the same commitment that is shown in stopping new invasions should also be applied in finding solutions to current invasions. For, example, can we use plasma solutions on zebra mussel infestations in the great lakes by demarcating certain areas and treating them with this application or more effective ones? This is a question that innovators and research and development can answer but the encouragement of these ideas is called for by this paper.

5. Ballast water in the Arctic region

The Arctic Region is regulated by the Polar Code developed by the IMO [29]. According to the Polar Code, The Ballast Water Convention (BWM Convention, 2004) and its Regulations D-1, regarding ballast water exchange (a method that requires foreign ballast water carried on a ship to be exchanged with local water before the ship enters the region's marine environment), an exercise that may prove dangerous in adverse weather but accepted to be effective and an acceptable practice, and D-2, regarding ballast water performance standards, are methods currently considered to be appropriate for the Arctic region [30]. The fact that Polar Code so clearly encourages compliance with the treatment of Ballast Water in that region by providing as follows:

"4 **Additional Guidance Under Other Environmental Conventions And Guidelines** 4.1 Until the International Convention for the Control and Management of Ships' Ballast Water and Sediments enters into force, the ballast water management provisions of the ballast water exchange standard, set out in regulation D-1, or the ballast water performance standard, set out in regulation D-2 of the Convention should be considered as appropriate. The provisions of the Guidelines for ballast water exchange in the Antarctic treaty area (resolution MEPC.163 (56)) should be taken into consideration along with other relevant guidelines developed by the Organization. 4.2 In selecting the ballast water management system, attention should be paid to limiting conditions specified in the appendix of the Type Approval Certificate and the temperature under which the system has been tested, in order to ensure its suitability and effectiveness in polar waters."

The Polar Code is a strong risk management and safety instrument mandated and applied together with the Safety of Life at Sea (SOLAS) [31] IMO instrument as well as the environmental protection law in the form of The International Convention for the Prevention of Pollution from Ships (*MARPOL*) [32]. This means that ballast water laws must be effective in the Arctic to protect that region (**Figure 2**).

In a paper, simulating ballast mater movements and management in the Arctic region authors Rosenhaim et al. use a ballast water tracer as a pathway to highlight through their experiments that ballast water accumulation in the Arctic is strongly linked to seasonality, this essential data for environmental protection and planning. Through their experiments, the authors demonstrate that conditions linked to summer, winter, and autumn conditions affect ballast water accumulation in the region. Most importantly, the authors through experimentation show that there is a risk of contamination of the environment through ballast water accumulation [11]. Here is a ballast water management warning from the experiment of the authors in the Arctic:

"In winter, due to the small number of vessels navigating the Northeast Passage, the amount of ballast water tracer released in the model was small. Following the increase in the number of vessels towards spring and especially in summer, the amount of ballast water tracer increased, and thus the risk of environmental contamination (e.g., by nonindigenous species, anthropogenic contaminants, pathogens, and toxins) [11]."

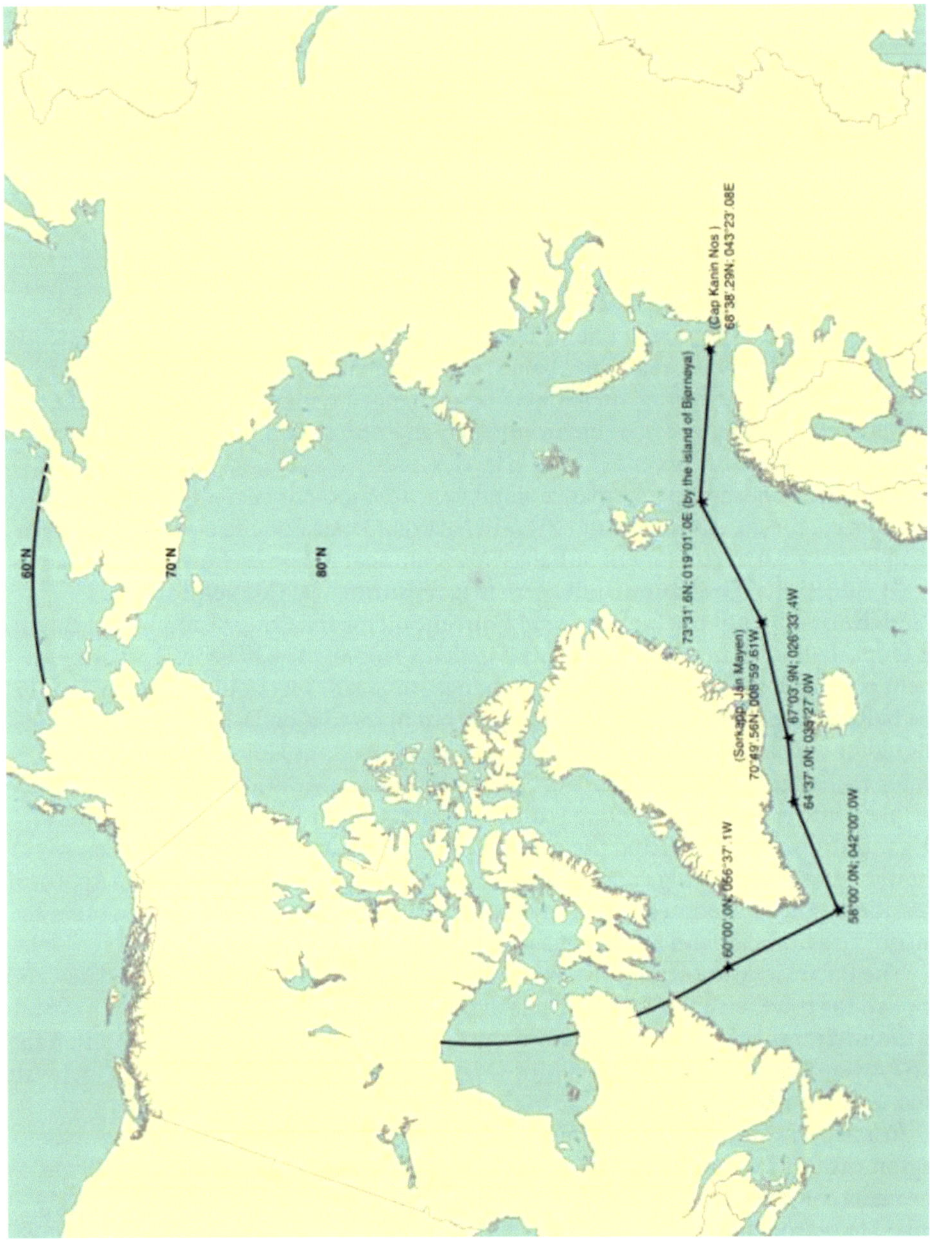

Figure 2.
Extent of Arctic waters as per IMO polar code, source: (IMO, polar code, 2017).

The authors above are telling us that when the Arctic waters open up more as a result of trade, ballast water methods on ships should be updated to comply with the BWM 2004 Convention, so that environmental contamination is a smaller risk for winter months (trades and voyages significantly shrink), or whether the contamination is more significant in the summer months, their tracer experiments should be able to (with the right ballast water technologies on board vessels trading in the Arctic) show that in general employing such technologies actually through the proof of data decreases contamination through ballast water. This is a powerful experiment to test for the success of ballast water management technologies.

6. The loss of Russian scientists in Arctic research

Protection of the environment is a global effort, where experts through effective and established and hopefully ethical methods put forth efforts to collaborate on sharing information on keeping environments safe whether it is space, land, or sea. This is why it is concerning that reports such as, "Russian and Western scientists no longer collaborate in the Arctic" [33] are quite disturbing. This situation is one of the direct results of the Russia-Ukraine conflict. It is submitted that from an international perspective policies ought to be developed concerning matters where the common interests of mankind are not abandoned to the extent of ignoring environmental protection since it is in the immediate interest of all, whether in conflict or not to continue collaborating on threat prevention efforts to protect the environment. For conflicting nations, basic survival takes precedence over all other interests that are understood however it is submitted in this paper that such situations should be treated in a special manner at the international level.

The maritime world is a global, interlinked one therefore having considered the impact of war on scientific minds, it is essential that we briefly consider ballast water management and impacts of other trade regions as a way of random global sampling in order to have a clearer perspective on some of the latest issues around ballast water management in the world. We will be considering Ballast Water management around Antarctica and the ancient trade route of the Mediterranean Sea, all areas where scientific minds globally have contributed.

6.1 Ballast water in the Antarctic region

It is important to note that the BWM 2004 Convention applies to the Antarctic region. There is also an important document in the form of the Antarctic Treaty, 1959 which guides the use, exploration and exploitation of the Antarctic. While dynamics of Antarctica are different from those of the Arctic in that the Antarctic is not a populous area with attractive shipping lanes for trade. However, due to untreated ballast water being introduced in the area, researchers also warn of the dangers of foreign waters being introduced together with AIS into the Antarctic Treaty region [34]. Authors, Dulière et al. in a scientific experiment running along a 9-year period conducted along the location of the Western Antarctic Peninsula, evidence of the introduction of AIS from ballast water can be overcome by encouraging ships to exchange foreign ballast water at least 200 nautical miles from the Antarctic region. This is the observance of the D1 standard of the BWM 2004 Treaty. This clearly means that the risk is extremely serious for the environmental future of the region and it will be essential for international parties to tighten sanctions for offending parties who traverse this region and pollute with impunity.

6.2 Ballast water in the ancient trade route of the Mediterranean Sea region

In a cutting-edge, very first study of its kind in the Mediterranean Sea conducted by Matej et al. [35], a sampling of 15 ships calling at the Port of Koper, Slovenia showed that ballast water management and the sometimes irreversible effects and damage on the environment are still very serious issues, particularly for a trade rouse as busy as the Mediterranean Sea shipping lanes. The results of the experiment showed that invasive species were being introduced with shipping as a vector, however, this was prior to the compliance and technological methods introduced by the coming into force of the BWM 2004. As part of ongoing research and monitoring, it would be very interesting what a repeat of this experiment might produce following the coming into force of the ballast water treaty.

7. Monitoring in environmental protection efforts

One of the concerns as to the effectiveness of ballast water management efforts and other environmental protection methods is the issue of monitoring. Monitoring helps in establishing a record of the success of the environmental protection innovations that we are using. In terms of innovation and technological developments, there have been impressive monitoring systems by various innovators. In this paper, the author highlights systems for monitoring by CEMTEK [36]. Without proof of data for compliance, even showing and measuring regulatory compliance is difficult. It is therefore proposed that ballast water management systems should also be equipped with tried and tested monitoring techs that will make it easier to reach standards and regulatory compliance while supporting ship management efficiency and protecting the environment.

8. Conclusion

Ballast water management has created excellent results in encouraging research and development as well as international cooperation on environmental protection to keep bio-invasions in the marine environment at bay. However, more data needs to be collected in order to measure the true effectiveness and human/biosafety of the more popular and now existing technological innovations to treat ballast water. The use of chlorine on the scale that it is being used to treat ballast water and the concerns raised about it must encourage even better solutions. This is true progress.

Further, taking special care for the ballast water management of a sensitive regions like the Arctic, the Antarctic, and the ancient traditional routes of trade around the world such as the Mediterranean Sea will need to continue. Collaborations to use science to create sustainable and bio-safe maritime activity must be the inspirational goal through which scientific studies for innovation, monitoring and record-keeping must be done while protecting sustainable maritime commerce.

The best and most environmentally sound ballast water management systems must be encouraged for that region to protect the environment and directly the communities of that region. In terms of the loss of Russian scientists because of the Russian-Ukraine conflict, cutting-edge research, ground-breaking findings of scientists should not be so easily abandoned, negotiation, and respect for scientific answers should still be protected, conflicting nations also have a vested interest in the environment and this must be protected. Finally, since the surveys conducted and experiments discussed in this paper have demonstrated where the strengths and weaknesses are (for example, consider the experiments in the Arctic and Antarctica, there are still problems with lack of compliance with the BWM 2004 Treaty that seriously threaten those environments), this should be very useful to guide different budgets to be prepared by governments, business entities, and various commercial interests to invest in making sure that ballast water management is taken very seriously.

Acknowledgements

I would like to thank IntecOpen for allowing the author to share their passion for research, education, and applied knowledge.

Notes/thanks/other declarations

To the future of the environment and all the children of the earth that are supported by it. A special thanks to the company representatives who granted me interviews to learn more about their innovations. The author declares that these interviews were research-based and the author does not represent the companies. The information shared here is also in the public domain herein synthesized to give educational, innovation inspiration and regulatory compliance tools for readers.

Author details

Fikile Portia Ndlovu
Massachusetts Maritime Academy, Buzzards Bay, United States of America

*Address all correspondence to: pndlovu@maritime.edu

References

[1] Freedom Ballast. Barge Based/Port Side, Ballast Water Treatment as a Service [Internet]. 2022. Available from: http://freedomballast.com/ [Accessed: April 4, 2022]

[2] International Maritime Organization. Ballast Water management, The Control of Harmful Invasive Species [Internet]. 2022. Available from: https://www.imo.org/en/MediaCentre/HotTopics/Pages/BWM-default.aspx [Accessed: April 4, 2022]

[3] Bailey SA. An overview of thirty years of research on ballast water as a vector for aquatic invasive species to freshwater and marine environments. Aquatic Ecosystem Health and Management. 2015;**18**(3):261-268. DOI: 10.1080/14634988.2015.102712

[4] International Maritime Organization Head Quarters (IMOHQ). Invaders from the Sea, a BBC-Worldwide Production [Internet]. 2022. Available from: https://youtu.be/u5JkRtMTEdI [Accessed: April 7, 2022]

[5] International Maritime Organization Head Quarters (IMO). Ballast Water Management (BWM) Convention. [Internet]. 2022. Available from: https://www.imo.org/en/OurWork/Environment/Pages/BWMConventionandGuidelines.aspx [Accessed: June 15, 2022]

[6] Ndlovu FP. The marine environment and ballast water management law. Water Policy. 2016;**18**(4):964-982

[7] Mark EC, Sweatman HPA, Brainard RE. The 2014-2017 global-scale coral bleaching event: Insights and impacts. Coral Reefs. 2019;**38**(4):539-545

[8] Naar JP, Flewelling LJ, Abbott JP, et al. Brevetoxicosis red tides and marine mammal mortalities. Nature. 2005;**435**(7043):755-756

[9] Merchant ND, Pirotta E, Barton TR, et al. Monitoring ship noise to assess the impact of coastal developments on marine mammals. Marine Pollution Bulletin. 2014;**78**(1-2):85-95

[10] Gudmestad OT. Technical and economic challenges for Arctic coastal settlements due to melting of ice and permafrost in the Arctic. In: IOP Conference Series. Earth and Environmental Science. Vol. 612, No. 1. Witpress; 2020. p. 12049

[11] Rosenhaim IL, Riemann-Campe K, Sumata H, et al. Simulated ballast water accumulation along Arctic shipping routes. Marine Policy. 2019;**103**:9-18

[12] IMO. International Convention for the Control and Management of Ships' Ballast Water and Sediments (the Ballast Water Management Convention, 2004). The International Maritime Organization; 2004

[13] International Maritime Organization. Latest Table Lists of IMO Approved Systems [Internet]. 2022. Available from: https://wwwcdn.imo.org/localresources/en/OurWork/Environment/Documents/Table%20of%20BA%20FA%20and%20TA,%20updated%20December%202021.pdf [Accessed: June 7, 2022]

[14] International Maritime Organization. International Convention for the Control and Management of Ships' Ballast Water and Sediments (the Ballast Water Management Convention, 2004). IMO D2 Standard for Ballast Water Management Technology on Ships: BWM Convention. The International Maritime Organization; 2004

[15] Darity WA. Likert scale. In: International Encyclopedia of the Social

Sciences (0-02-865965-1, 978-0-02-865965-7). Macmillan Social Science Library; 2008. p. 447

[16] Ndlovu FP. Industry Attitudes: Survey Results and Comments. Fikile Portia Ndlovu, Newcastle upon-Tyne: Cambridge Scholars Publishing; 2022. pp. 112-130

[17] Elkady H, Feng Han D, Gao Gao L. The alternatives of ballast water systems. Journal of Applied Mechanics and Materials. 2014;**627**:347-352

[18] Interview with Mpumelelo Sehoane, Transnet Marine Engineer; 2022

[19] Illustration by Miami Shark Research, article by Dresbach C. Ballast Water Management and Its Implications Regarding Invasive species Introduction. 2016. Available from: https://sharkresearch.rsmas.miami.edu/ballast-water-management-and-its-implications-regarding-invasive-species-introduction/ [Accessed: April 7, 2022]

[20] Schuster LAA. Workforce Divided: Community, Labour and the State in Saint Nazaire's Shipbuilding Industry 1880-1910. Westport, CT: Greenwood Press; 2002. p. 81

[21] Johnson C. Seaweed Invasions: A Synthesis of Ecological, Economic and Legal Imperatives. Berlin, Germany: Walter de Gruyter GmbH & Co.; 2008. pp. 113-433

[22] McConnell M. GloBallast Legislative Review–Final Report. GloBallast Monograph Series No. 1. Vol. 71. London: IMO; 2002. pp. 45-47

[23] Waldenmaier RW. Approved ballast water treatment systems. In: Ballast Water Management and Environmental Protection. Fikile Portia Ndlovu, Newcastle upon-Tyne: Cambridge Scholars Publishing; 2022. pp. 147-149

[24] Dececca A, Foley H. "Ballast water management technologies and solutions: Projecting and comparing: A snapshot analysis" in Ballast Water Management and Environmental Protection, Ed. Fikile Portia Ndlovu, Newcastle upon-Tyne: Cambridge Scholars Publishing; 2022. pp. 80-97

[25] [Internet]. 2022. Could Your Ballast Water Treatment Be Killing You? Available from: https://www.martek-marine.com/blog/could-your-ballast-water-treatment-be-killing-you/ [Accessed: June 7, 2022]

[26] D-2 – IMO STANDARD: 'Ballast water performance standard D2

[27] Lee K-J, Shin D, Yoo D-W, et al. Hybrid photovoltaic/diesel green ship operating in standalone and grid-connected mode–experimental investigation. Energy (Oxford). 2013;**49**:475-483

[28] Freedom Ballast. Barge Based/Port Side Ballast Water Treatment as A Service [Internet]. 1999. Available from: http://freedomballast.com/ [Accessed: March 29, 2022]

[29] International Code for Ships Operating in polar Waters (The Polar Code), 2017

[30] International Code for Ships Operating in polar Waters (The Polar Code). See part 4 of additional guidance on the code on environmental considerations. In: Pandey A, editor. Ballast Water, a Case for Arctic Shipping. Vol. 2022. Fikile Portia Ndlovu, Newcastle upon-Tyne: Ballast Water Management and Environmental Protection; 2017. pp. 181-182

[31] IMO. International Convention for the Safety of Life at Sea (SOLAS). The International Maritime Organization; 1974

[32] IMO. International Convention for the Prevention of Pollution from Ships (MARPOL). The International Maritime Organization; 2022

[33] The Economist. Russian and Western scientists no longer collaborate in the Arctic [Internet]. 2022. Available from: https://www.economist.com/science-and-technology/russian-and-western-scientists-no-longer-collaborate-in-the-arctic/21809236 [Accessed: June 7, 2022]

[34] Dulière V, Guillaumot C, Lacroix G, et al. Dispersal models alert on the risk of non-native species introduction by ballast water in protected areas from the Western Antarctic peninsula. Diversity & Distributions. 2022;**28**(4):649-656

[35] David M, Gollasch S, Cabrini M, Perkovič M, Bošnjak D, Virgilio D. Results from the first ballast water sampling study in the Mediterranean Sea – The port of Koper study. Marine Pollution Bulletin. 2007;**54**(1):53-65. DOI: 10.1016/j.marpolbul.2006.08.041

[36] CEMTEK. [Internet]. 2022. Available from: https://cemteks.com/ [Accessed: April 4, 2022]

Section 3

Marine Pollution with Microplastics

Chapter 6

Microplastics in the Marine Environment: A Review of Their Sources, Formation, Fate, and Ecotoxicological Impact

Fatima Haque and Chihhao Fan

Abstract

Global plastic production is on the rise, and improper plastic management leads to the disposal of plastic in the environment, wherein it enters the environment, after degradation, as microplastics (size < 5 mm) and nanoplastics (size < 1 μm). The most common sink for the microplastics is the marine environment, including the sediment, deep sea, shorelines, and oceans. The objective of this study is to collate the environmental impact assessment of the microplastics in the marine habitat, focusing on the following main elements: (a) source and type of microplastics, specifically leading to the marine sink; (b) degradation pathways; (c) ecotoxicological impact on marine biota, since the smaller-sized microplastics can be digested by the marine biota and cause threats to them; (d) fate of microplastic in the marine environment, including the modes of transport and deposition. This chapter aims to provide a deeper insight into the fate of microplastics once it enters the marine environment, and the information could be a useful reference for the development of microplastic risk management strategies.

Keywords: microplastics, plastic waste, marine habitat, ecotoxicology, degradation

1. Introduction

Global plastic production is on the rise wherein 1.3 million tons of plastics were produced in 1950, and 359 million tons of plastic waste were generated in 2018 [1, 2]. It is estimated that the increase in plastic waste will reach 250 million metric tons by 2025 [3]. This adds additional pressure on the plastic management system. At present, 9% of the plastic waste is recycled, 50% ends up in landfills, 19% is incinerated, and the remaining 22% ends up being discarded as litter (and is categorized as mismanaged plastic waste) [4, 5]. The mismanaged plastic waste is often dumped on terrestrial lands or in marine habitats [6]. It has been estimated that 10% of the mismanaged plastic waste ends up in the marine environment where it will persist and accumulate over the coming years [7]. The large fragments of plastic debris found in the environment are termed macroplastics [8], and they are known to harm turtles and sea birds via entanglement [9, 10].

Once these macroplastics enter the environment, they undergo degradation and decompose into smaller fragments known as microplastics (size < 5 mm) and/or nanoplastics (size < 1 μm) [11, 12]. Microplastics can be differentiated into primary microplastics and secondary microplastics, depending on their sources. Primary microplastics are the ones manufactured for direct applications such as microbeads in personal skin care products [13–15]. Secondary microplastics are the ones formed as a result of the degradation and decomposition of the macroplastics [16]. The most common sink for microplastics is the marine environment, including the sediment, deep sea [17, 18], shorelines [19, 20], oceans [21, 22], and interestingly coral reefs as well [23].

Globally, microplastics are recognized as pollutants, and the United Nations Sustainable Development Goals (UN SDG) has assigned Goal 14 specifically to conserve and sustainably use the oceans, seas, and marine resources [24]. The contamination by microplastics and nanoplastics has been an issue of concern over the past decade. Owing to their small size, micro/nano plastics are readily bioavailable for consumption by marine organisms [25]. Once ingested by smaller marine organisms (primary consumers), they will be further transferred to the secondary consumers (e.g., large fishes) and eventually reach the tertiary consumers (e.g., humans), thus disrupting the food chain [26].

Though the sources, degradation pathways, and sinks (specifically marine habitat) of the microplastics are often discussed, the fate of microplastics is elusive after perusing various articles and literature. Through this chapter, we aim to collate the environmental impact assessment of the microplastics in the marine habitat, focusing on the following main elements: (a) sources of microplastics, their transport to the marine environment, as well as their types; (b) degradation pathways including photodegradation, weathering, corrosion, or mechanical forces of water; (c) ecotoxicological impact on marine biota, since the fragmented microplastics can be readily digested by the marine biota and cause a threat to them; (d) fate of microplastic in the marine environment, including the modes of transport and deposition. This chapter aims to provide a platform for the development of microplastic risk management strategies and also to provide a deeper insight into the fate of microplastics once it enters the marine environment.

2. Sources, transport, and type of microplastics

In this section, we examine the main sources of microplastics, followed by how they reach the marine environment. Lastly, the types of microplastics predominant in the marine ecosystem are summarized.

2.1 Sources of microplastics

The sources of microplastics can be categorized into primary and secondary sources, and each category is discussed as follows.

2.1.1 Primary sources

Primary sources of microplastics include: **plastic pellets**, also known as nibs (diameter: 2–5 mm), which are used to make various types of plastic products [27];

microbeads, which are used in the manufacturing of personal care products, face wash, face cleansers, facial scrubs, hair products, nail polish, deodorants, sunscreen, and eye shadows [13–15]; **glitters,** which are shiny substances found in cosmetics and textile products. They are usually made of polyethylene terephthalate (PET) polymer, acrylic, polyvinyl chloride (PVC), and/or polymethyl methacrylate (PMMA) [28]. These primary plastics vary in shape, size, and composition depending upon their applications [15]. For example, certain cosmetic products contain granules of polyethylene and polypropylene (<5 mm), spheres of polystyrene (<2 mm) [29], or irregularly shaped microplastics (<0.5 mm) [15]. Apart from cosmetics, these primary sources of microplastics also find applications in air-blasting technology [14, 29]. This technology uses acrylic, melamine, or polyester as scrubbers at high pressure on machines, engines, and water vessel hulls to scrape off rust buildup or paint [13, 30].

2.1.2 Secondary sources

2.1.2.1 Effluent from water and wastewater treatment plants

Water and wastewater treatment plants are one of the main sources of releasing microplastics into the marine environment [31]. They are found in the primary stages of water treatment. Because of their small size, they can pass through the filters and enter the secondary units [32]. Microplastics detected in the influents ranged from ~1 to 10,000 particles per liter, and after treatment, microplastics in the effluent ranged from ~0 to 450 particles per liter (as summarized by a number of studies reviewed by Sun et al. [33]). Microfibers, including polyester, acrylic, and polyamide, are detected in the effluent of wastewater treatment plants [34], which implies the limitations of these treatment facilities to remove these microplastics.

2.1.2.2 Wear and tear from normal plastic use

The most common example of such a source type is the microplastic released as a result of washing clothes and textiles during laundry [35]. As a result, microplastics released from laundry activities eventually reach the marine environment. It is estimated that laundry activities are responsible for 500,000 tons of microplastics in the ocean per year [36, 37]. Apart from textiles/clothes weathering, use of fishing gears, including nets and ropes [38], wear and tear of car tires [39], as well as weathering of household items, including toys, plastics wares, and plastic disposables items [40].

2.1.2.3 Airborne dust

Plastic dust is released from a number of activities including plastic manufacturing facilities, incineration of plastic wastes, traffic emissions, weathering of roads and streets, and urban mining activities [41, 42]. Airborne dust is carried by wind and can settle in indoor settings including schools and houses [43, 44]. In houses, airborne microplastic comes from plastic items used in household items including food packaging, plastic wear, and plastic furnishings [45]. Most recently, during the COVID-19 pandemic, the requirement to wear face masks was made mandatory to prevent the spread of coronavirus. The surgical facemasks were made up of PP, PE, PS, and polyester. Studies showed that wearing these masks exposed the humans directly to inhalation of micro (<1 μm) and nanofibers (<100 nm) [46–48].

2.1.2.4 Secondary microplastics

Primary microplastics may also contribute to the secondary sources of microplastics. Once exposed to the environment, plastic wastes and primary sources of microplastics undergoes weathering and degradation to form secondary microplastics [12]. Details on the degradation process of plastic waste are given in Section 3. Plastic litters including disposable plastic cutlery, plastic cups, food containers, as well as face masks in the era of COVID-19 pandemic (that started in 2019 and is still ongoing in the current year of 2022) end up being dumped on coastal shorelines, where they undergo further degradation and decomposition [48–50].

2.2 Transport

There are four main pathways through which microplastics from different sources reach the marine environment: (a) as surface runoff when the plastic wastes are thrown on the terrestrial lands and eventually travel along with the runoff due to rainfall. Transport via surface runoff is responsible for 44% of the total microplastics being released into the marine ecosystem; (b) via wind, which transports the plastic waste on the terrestrial zone to seas/oceans along with the atmospheric currents. Transport via wind is responsible for 15% of the total microplastics being released into the marine ecosystem; (c) as wastewater discharge in which microplastics can enter the receiving water bodies and is responsible for 37% of the total microplastics released; (d) and lastly, through direct disposal of plastic wastes into the marine environment, which is responsible for 4% of the total microplastics release [7, 13, 30, 51]. Direct disposal of plastic wastes activities includes washing clothes in the rivers, usually in the rural areas [52], coastal tourism activities including fishing and recreational activities resulting in disposable cups and litters [53], and commercial fishing resulting in nets and litters [54].

2.3 Types of microplastics

Microplastics can be categorized into primary and secondary microplastics depending on their sources, as discussed in Sections 1 and 2.1. Depending on their density and chemical compositions, microplastics can be classified into different types including polystyrene (PS), low-density polyethylene (LDPE), high-density polyethylene (HDPE), polypropylene (PP), polyvinyl chloride (PVC), polyethylene terephthalate (PET), and others (e.g., nylon, polyester) [55, 56]. The different plastic types, properties, and functions where these plastics are commonly used are given in **Table 1**. Microplastics can also be differentiated on the basis of shape: pellets, microbeads, foams, fibers, films, fragments, and microfibers (**Figure 1**) [57].

3. Degradation pathways

Plastic wastes undergo environmental weathering resulting in the formation of microplastics or even smaller fragments of nanoplastics. These degradation pathways can be classified into abiotic and biotic processes [62–64].

Plastic type	Abbreviation	Properties	Common applications
Polystyrene	PS	Density (1.04–1.08 g/cm^3) transparent, hard.	Personal care products (as microbeads), household items (utensils and containers), disposable cups, plastic components of electronic instruments, and packaging.
Low-density polyethylene	LDPE	Density (0.89–0.94 g/cm^3), translucent, soft.	Clingy plastic wraps and films, containers, plastic bags, and flexible pipes and tubing.
High-density polyethylene	HDPE	Density (0.94–0.97 g/cm^3), opaque, hard/semi-flexible.	Food packaging (cereal box liners, milk bottles), freezer bags, plastic stools, courier envelopes, and toys.
Polypropylene	PP	Density (0.89–0.91 g/cm^3), translucent, hard.	Straws, packaging tapes, snack bags (chips and biscuit bags), fishing gears (nets and ropes), bottle caps, clothing, textiles, and microbeads in skin care products.
Polyvinyl chloride	PVC	Density (1.3–1.58 g/cm^3), transparent (clear), hard.	Medical supplies (blood bags, surgical gloves and face masks), building structures (floorings, roof plates, swimming tanks, and fittings), shoes, and tents.
Polyethylene terephthalate	PET	Density (1.29–1.4 g/cm^3), transparent, hard.	Food packaging (clamshell packaging in takeaway containers such as salad domes, biscuits and snack trays), thermal insulations, and textiles.
Others	Ex.:polyester, polyamide (nylon)	Density of polyester (1.01–1.46 g/cm^3), Density of polyamide (1.13–1.35 g/cm^3).	Packaging, nylon products, textiles, abrasives in cleaning supplies.

Table 1.
Properties and common applications of different types of plastic found in the marine environment [55, 56, 58–61].

3.1 Abiotic degradation pathway

Abiotic factors include mechanical forces that are responsible to damage the plastic wastes physically, temperature increase (thermal degradation), chemical degradation, and light irradiation (leading to photodegradation) [12].

3.1.1 Mechanical degradation

Mechanical degradation refers to the action of external forces caused by wind current, ocean waves, or physical wear and tear resulting in breakdowns of plastics [12]. Plastic litters on coastal shorelines are exposed to collision and abrasion with beach rocks and sands as a consequence of motion caused by wind and ocean circulations. In the colder zones, repetitive freezing and thawing of ice can cause the degradation of the plastics accumulated in the ice and eventually result in their flow back into the marine habitat [65, 66]. One example of mechanical degradation of plastic is wear

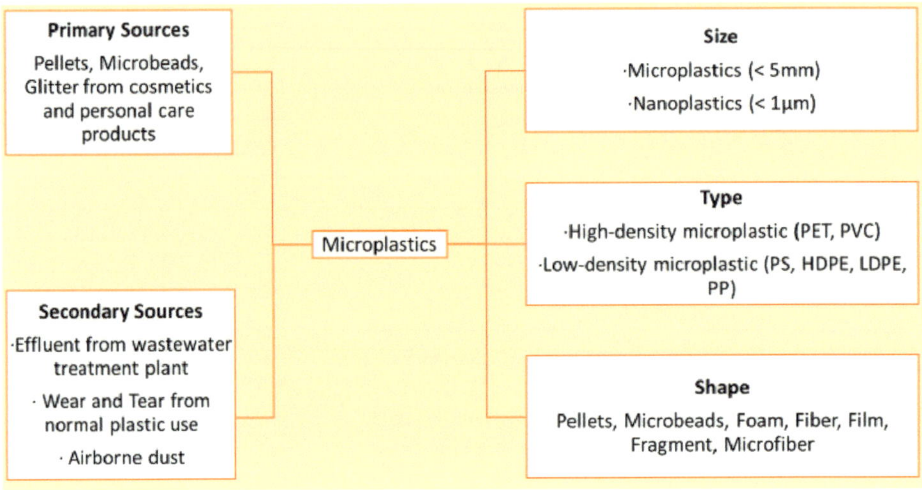

Figure 1.
Characterization of microplastics based on source, size, type, and shape.

and tear on the road as a result of friction caused by the moving car tires [67]. Tire, road, and brake wear happens because of the mechanical forces being exerted on the brake pads, tire threads, and the road surface, resulting in material stressing and fatigue [68].

3.1.2 Thermal degradation

Plastic waste litters on the coastal shorelines are exposed to elevated temperatures, leading to a thermo-oxidative breakdown of the plastic. Thermal degradation of plastics involves absorption of heat and breaking of polymeric chains thus releasing radicals that react with atmospheric oxygen to produce hydroperoxide, which eventually cleaves into hydroxyl and alkoxy free radicals. These radicals result in the formation of aldehydes, ketones, esters, or alcohols, causing plastic degradation [69]. Chain scission and cross-linking of the polymers are responsible for the thermal degradation process [70, 71]. In the environmental matrix related to beaches and coastal shorelines, slow thermal degradation of plastics may occur concurrently with photodegradation (due to the presence of sunlight), resulting in enhanced plastic degradation [72].

3.1.3 Chemical degradation

Chemical pollutants are present in the atmosphere (e.g., sulfur dioxide, nitrogen dioxide, ozone, and volatile organic compounds) and the marine environment (e.g., acidity and salinity). The atmospheric pollutants can directly degrade the plastics or catalyze the radical formation by photochemical reactions leading to plastic degradation [73]. Sulfur dioxide and nitrogen dioxide can enhance the formation of ozone in the atmosphere, as a result of UV excitation and photochemical reaction with oxygen [74]. The ozone formed can break the carbon double bonds present in the plastic polymers (chain scission mechanism). In the marine environment, the acidity or alkalinity of the water can catalyze plastic degradation such as polyamides [75].

3.1.4 Photodegradation

Photodegradation of plastic is mediated by sunlight UV radiations, both UVB (290–315 nm, high-energy radiation) and UVA (315–400 nm, medium-energy radiation) [12, 76]. Photodegradation of plastic involves free radical formation and oxidation of the plastic polymers, resulting in the formation of peroxides, which eventually breaks into alkoxy and hydroxyl radicals, similar to the thermal degradation mechanism. Photodegradation in the atmosphere results in the formation of free radicals to break different plastics depending on their chemical structures. For example, the presence of chromophores (alternating or conjugating carbon double bonds) in PP, PE, and PVC, phenyl rings in PS, and ethylene glycolate and terephthalate groups linked with ester bonds in PET mediate the free radical formation reactions as a result of photodegradation [12].

3.2 Biotic degradation pathway

Plastic degradation by microorganisms present in the marine habitat results in the biodegradation of plastic wastes. However, macroplastics (larger plastic debris) are not the ideal feedstock for biotic degrading agents owing to their size, which poses a hindrance to the degradation mechanism, either the enzymes produced by the microorganisms are not enough to degrade the macroplastics, or they are not readily bioavailable for microbial cell uptake. During the degradation process, polymeric plastics need to be first converted into monomers before they can be mineralized by the biological agents. The molecular size of plastics (i.e., polymers) is larger than the pore size of microorganism's cell membrane. Hence, they need to be depolymerized into smaller fragments before they can be absorbed and biodegraded within the microbial cells. Therefore, smaller fragments of plastic formed as a result of abiotic degradation are of the appropriate size to be further degraded by microorganisms [12]. Microorganisms predominantly present in the marine environment include bacteria, fungi, and algae.

3.2.1 Bacteria

Bacillus species are commonly found in the marine environment, for example, *Bacillus subtilis* and *Bacillus cereus.* These bacteria were found to secrete extracellular hydrolytic enzymes such as *lipase, xylanase, keratinase, chitinase, and protease*, which lead to plastic degradation [77]. PVC, the most common plastic polymer, can be degraded by *Methanosarcina barkei*. They can adhere to the surface of the PVC surfaces and release exopolymeric substances to form a biofilm on the PVC, followed by the release of enzymes to degrade the plastic via hydrolytic cleavage of the polymeric bonds [78, 79]. Similarly, PE can be degraded by *Rhodococcus ruber*, which produces an enzyme laccase that results in PE degradation [80]. PS can be degraded by *Azotobacter spp.*, which produces hydroquinone peroxidase. PET can be degraded by *Alcanivorax*, *Hyphomonas,* and *Cycloclasticus* species, which can change the surface chemistry via hydrolysis of the ester bonds [81].

3.2.2 Fungi

Fungi can also result in biotic degradation of plastics. For example, *Aspergillus clavatus* has been shown to biodegrade LDPE [82]. Oceans' predominant fungal species *Zalerion maritimum* can degrade PE [83]. Similar to bacteria, the main mechanism of plastic degradation by fungi involves the adherence of the fungi to the plastic surface, where they grow to form a biofilm and produce enzymes to break down the chemical bonds present in the plastic. These enzymes can catalyze oxidation-reduction reactions and break down plastic into smaller fragments (e.g., oligomers, dimers, and monomers). For example, manganese peroxidase, lignin peroxidase, and laccase are produced by fungi present in marine habitats, such as *Penicillium citrinum* (degrades PET), *Fusarium oxysporum* (degrades PET), and *Trichoderma harzianum* (degrades PE and PU) [83].

3.2.3 Algae

Some algae have been shown to produce secondary metabolites that can biodegrade microplastics. For example, *Phormidium lucidum* and *Oscillatoria subbrevis* can biodegrade PE and LDPE [84]. Algal biofilms formed by *Discostella spp.*, *Navicula spp.*, *Amphora spp.*, and *Fragilaria spp.* have shown to degrade LDPE, PP, and PET in the marine environment [85]. Once forming a biofilm on the plastic surface, algae utilize the carbon present on the plastic as a source of nutrition, thus weakening the strength of the plastic and making it fragile. Moreover, algae produce extracellular polymeric substances and enzymes such as PETase that result in the degradation of PET [86]. Plastic degradation by algae is still in its nascent phase and needs further research.

4. Distribution and fate of microplastic

The distribution and fate of the degraded plastic and microplastic in the marine system are attributed to anthropogenic activities (e.g., tourism, wastewater treatment effluent) in the form of primary microplastics [87]. Environmental factors lead to the introduction of secondary microplastics into the marine habitat, as discussed in Section 3. For example, the wastewater treatment plant effluent releases ~7 million microplastic particles every day [87, 88]. Hence, the marine environment serves as the primary sink for microplastics. Once into the marine system, their accumulation and distribution depend on a number of parameters pertaining to microplastics (e.g., density, size, shape, and chemical composition) and environment (e.g., wind and ocean current speed) [89, 90]. The fate of microplastics is related to their immediate source of disposal, and they can be translocated to remote areas such as artic seas and ice-capped regions [91]. Depending on the density of the microplastics, they can either remain suspended in the surface water or sink into the deep sediments. The density and other chemical properties of the most common types of microplastics are given in **Table 1**. If the density of the microplastic is lesser than that of the seawater (usually ~1.025 g/cm^3) [61], the microplastic may remain suspended in surface water and would be transported to distant locations through horizontal distribution driven by ocean circulations (Section 4.1). If the density of the microplastic is greater than that of the seawater, the microplastic may sink to the sea floor through a pathway of vertical distribution (Section 4.2) [61, 92, 93]. Data show that around 15% of microplastics remain in the suspended form, whereas 70% of microplastics accumulate in sea sediments [94]. In the United States, ~260 tons of PET are released from the used containers of personal care products alone, and this contributes to 25% of microplastics in the North Atlantic Ocean gyre [95]. Due to the variation in degradation mechanisms of different plastics, the continuous generation of plastic waste, and the dynamic nature of the environmental conditions (since the velocities of wind and ocean circulation vary along with the changing weather conditions), the fate of microplastic is not constantly steady and difficult to predict. This necessitates a proper understanding of the distribution of the microplastics once it enters the marine system.

4.1 Horizontal distribution

Coastal current, rainfall, and wind are responsible for the movement of the plastics from the coastal shorelines/beaches into the marine system [96–98]. Once the macroplastics enter the marine environment, they can undergo further degradation as a result of ocean abrasion or biotic degradation, as discussed in Section 3. The fate of the microplastics, those carried from the terrestrial shorelines and/or formed as the result of degradation in the marine system, depends on their intrinsic properties and ambient conditions. Depending on the velocity and direction of flow of the regional wind and water current, these microplastics can either be transported to remote regions or return to the coastal shorelines/beaches [32, 87, 99], resulting in the accumulation of microplastics in the oceanic/regional water gyres in the marine environment. Meanwhile, 5–13 million tons of plastic debris enter the ocean (data for 2010) [3], and approximately 7–35 thousand tons of suspended microplastics remained in the ocean surface water [100]. This implies that the remaining plastic debris was translocated (either by horizontal or vertical distribution pathways). **Figure 2** shows the distribution pathways for plastic and microplastics in the marine environment.

4.2 Vertical distribution

As stated previously, microplastics with density greater than that of the marine/region water may sink to the seabed. This process is mediated by vertical turbulent mixing, biota transfer (via fishes or other marine organisms), biological fouling (also known as biofouling), and aggregate formation [61, 101]. Biofouling is the accumulation of existing marine microorganisms, planktons, algae, microalgae, and small marine organisms on the plastic debris/microplastics [102]. This process depends on the polymer type, surface area, and size of the microplastic, as well as the microorganisms present in the marine environment, temperature, salinity, pH, nutrient/metals, and oxygen concentration of the water [66, 103–106]. For example, the presence of a plethora of bacterial species (*Alteromonas, Zoogloea, Ruegeria, Roseobacter, Nautella, and Pseudomonas*) in the benthic (6 m in depth) and the planktonic (2 m in depth) zones of the Arabian Gulf resulted in the biofouling of PET and PE [107]. Another study showed that the water conditions, primarily oxygen concentration and the presence of iron in the water resulted in biofouling of PET, PE, and PS by cyanobacteria, bacteria, and algae [108]. Biofouling starts with the attachment of the organisms, nutrients, flocculants, and dissolved organic compounds on the microplastic surface [109]. Subsequently, extracellular polymeric substances are released by the microorganisms to form a biofilm, which further attracts other marine invertebrates and worms [110]. As a result, the aggregate forms, the overall density of the microplastic increases, and it eventually sinks.

The density of marine water varies at different depths. Therefore, depending on the density of the aggregate formed, different layers can serve as a sink to accumulate microplastics [102]. Heavier aggregates can sink into the deep oceanic layers. The fate of the microplastics accumulated in the marine sediments is affected by the

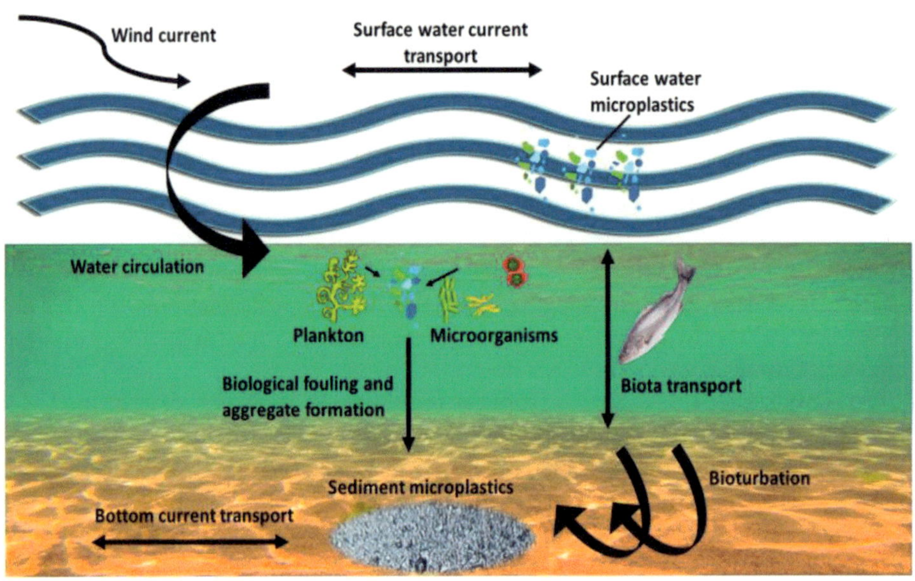

Figure 2.
Distribution pathways and fate of microplastics in the marine environment.

disturbance in the sediment zone, resulting in releasing the accumulated microplastics back into the water zone [111]. Also, similar to surface water currents, bottom water currents can also lead to the transportation of the microplastic to remote regions (**Figure 2**) [101].

5. Ecotoxicological impact on marine biota

Owing to their small size, microplastics have the potential to be ingested by an array of marine biota [112]. There are several studies indicating the ingestion and accumulation of microplastics in marine organisms, and most of the studies were conducted on fishes. **Table 2** lists a number of studies demonstrating the impact of microplastics on different marine organisms, categorized into fishes, invertebrates, and other miscellaneous biota. These studies indicated the accumulation of microplastics in various marine organisms including fishes (mackerel, *Scomber japonicus*), copepods (*Calanus helgolandicus*), and shorebirds (whimbrel, *Numenius phaeopus*), and pacific golden plover (*Pluvialis fulva*) [113–115]. When a microplastic accumulates in the organism's body tissues, it may influence the organism's health in numerous ways, including stunted growth, infertility, and impact on egg's hatching [114, 116]. Once ingested by the marine organisms, the microplastics can translocate through the food chain, starting from the primary consumers (e.g., planktons, small fishes), to the secondary (e.g., larger fishes, birds, turtles), and eventually to the tertiary ones (humans) [117]. Such a process is known as biomagnification, which may cause human health risks [32]. Moreover, microplastics can bind to various marine pollutants such as heavy metals, enhancing their accumulation in the marine environment [118]. In addition, marine invertebrates such as mollusks (e.g., mussels, oysters, clams) and crustaceans (e.g., shrimps, crabs, lobsters) do not possess the required digestive enzymes to break down the microplastics into simpler nontoxic compounds. Therefore, these invertebrates would release the microplastics back into the water as fecal matters [119]. As a result, the microplastic might not have any toxic impact on the marine organisms once the ingested microplastics are egested. In certain cases, microplastics can act as a vector of co-pollutants present in the marine system and prevent its translocation to the marine organisms, thus exhibiting a positive impact on the organism. For example, in the presence of co-pollutant (zinc oxide) and microplastics (PE), marine microalgae (*Dunaliella salina*) showed higher growth than in the absence of PE. This is because PE could attach to zinc oxide, leading to its leaching and preventing its uptake by the microalgae [120]. The ecotoxicological risk and impact of microplastics on the marine environment can be categorized into physical, chemical, and biological damages. Physical damage to marine organisms includes gastrointestinal tract blockage and damage, leading to the organism's death and affecting the mortality rate [121]. Chemical damage includes the property of microplastics acting as carriers or vectors for pollutants such as heavy metals (e.g., Cr, Ni, Cd, Zn) that are eventually ingested by marine organisms [122]. For instance, PE was found to facilitate the sorption of chromium (Cr) in common Goby fish, which led to a decrease in acetylcholinesterase (AchE) enzyme activity and resulted in acute toxicity [123]. Lastly, biological damage to marine organisms includes gene manipulation and the evolution of microorganisms with antibiotic resistance genes and metal resistance genes [124]. However, more research is needed to confirm the impact of these damages on marine organisms.

Table 2 also summarizes the ultimate marine sinks for the microplastics. The marine organisms impacted by the microplastics are primarily present in the following major oceans: the Pacific ocean, Atlantic ocean, and Indian ocean. Pacific ocean serves as a marine sink to microplastic generated from the United States (e.g., California [125]) and South America (e.g., Peru and Chile coastlines and Northern Patagonia in Chile [126, 127]). These examples represent the East Pacific Ocean as the marine sink for microplastics, where the main sources of these microplastics include plastic manufacturing industries in the United States (e.g., California [125]), and textile industries and domestic washing of clothing in South America (e.g., Peru and Chile [126, 127]). Likewise, the West Pacific Ocean serves as a microplastic source for marine habitats including zebrafish, rotifers, copepods, shrimps, scallops, crinoids, (China [128–130], gastropods, bivalves, and crabs (Hongkong [131]), as well as seabirds and turtles (China [115, 132]). Based on recent studies summarized in **Table 2**, the main source of microplastic in the West Pacific Ocean is China. The increased consumption of plastic in China is directly linked to its high population (1.41 billion [133]), plastic manufacturing industries, and mismanaged plastic wastes [134].

Similarly, increased fishing activities, tourism, and high population are the main reasons for the microplastic source of the Indian Ocean, including India (primarily the high population of 1.38 billion [133]) and Thailand (primarily the tourism activities and fishing activities, [135]). The source of microplastic to the Atlantic Ocean is due to the increased amount of plastic waste generated (e.g., United Kingdom), and the mismanaged plastic waste that makes it to the ocean [134]. For example, the East Atlantic Ocean serves as a microplastic source to marine organisms including common goby (Iberian coast, [123]), different pelagic and demersal species in the English Channel (UK, France [136–138]), gilthead seabream and European seabass (Murcia, Spain, [139]), mussels (Port Quinn Cornwall, UK, [140]), copepod (English Channel, UK, [114]), insects (Italy, [141]), whale (the Netherlands, [142]), and otters (Norway, [143]). Similarly, for the West Atlantic Ocean, the United States, and South America serve as a microplastic sink/source for different marine organisms including

Organisms	Sample location, major ocean sink	Type of MP	Impact	Reference
Fishes				
Commercial fish (26 species)	Portugal coast, Atlantic Ocean	PP, PE, polyester, nylon, acrylic, rayon, and resins.	*Scomber japonicus* ingested the highest amount of microplastics, mainly fibers and fragments.	Neves et al. [113]
Common goby (*Pomatoschistus microps*)	Estuaries of Minho River and Lima River (North-West Iberian Coast), Atlantic Ocean	PE	Presence of microplastic along with heavy metal chromium (Cr) resulted in decrease in acetylcholinesterase (AchE) activity. This results in acute toxicity of the fish towards Cr.	Luís et al. [123]
Zebrafish (*Danio rerio*)	Tianjin Baseline ChromTech Research Centre (Tianjin, China), Pacific Ocean	PS	Microplastic accumulated in the gills, guts and liver of Zebrafish. This resulted in multiple toxic effects including inflammation, increase in enzyme activity (superoxide dismutase and catalase). This leads to creating imbalance of metabolic pathways.	Lu et al. [128]
Japanese medaka (*Oryzias latipes*)	Aquatic Health Program at UC Davis (California), Pacific Ocean	PE	Ingestion of microplastic lead to disruption of normal functioning of the endocrine system. Down regulation in genes expression of choriogenin (ChgH) in male and vitellogenin (VTgI) & estrogen receptor (ERα) were reported.	Rochman et al. [125]

DOI: http://dx.doi.org/10.5772/intechopen.107896

Organisms	Sample location, major ocean sink	Type of MP	Impact	Reference
Five pelagic species (whiting *Merlangius merlangus*), blue whiting *Micromesistius poutassou,* Atlantic horse mackerel *Trachurus,* poor cod *Trisopterus minutus* and John Dory *Zeus faber)* Five demersal species (red gurnard *Aspitrigla cuculus,* Dragonet *Callionymus lyra*, redband fish *Cepola macrophthalma*, solenette *Buglossisium luteum*, and thickback sole *Microchirus variegates)*	English Channel (UK), Atlantic Ocean.	Polyamide, Rayon	37% of the fish examined (n = 504) had ingested MP, which causes mortality by choking or sub-lethal damage due to disruption of intestinal tissues.	Lusher et al. [136]
Silver barb (*Barbodes gonionotus)*	Malaysia, Indian Ocean.	PVC	During the first 4 days, there was no damage to the fish, but after prolonged exposure, intestinal damage occurred followed by increased trypsin and chymotrypsin activity.	Romano et al [151]
European sea bass (*Dicentrarchus labrax)*	Atlantic Ocean	PVC	Intestinal damage.	Peda et al. [152]
European sea bass (*D. labrax)* larvae	Marine farm Aquastream (France), Atlantic Ocean	PE	Injuries and ulceration in the intestines.	Mazurais et al. [137]
3 fish species (*Clupea harengus*, *Sardina pilchardus* and *Engraulis encrasicolus*)	English Channel, the Northwestern Mediterranean Sea and the Northeastern Atlantic (Bay of Biscay), Atlantic Ocean	PE, PP, PET	Reduced gill functioning	Collard et al. [138]
Goldfish (Carassius auratus)	Laboratory conditions	PS, PE	MPs were found in the gills, guts, and feces.	Jabeen et al. [148]

Organisms	Sample location, major ocean sink	Type of MP	Impact	Reference
European seabass (*D. labrax*), the Atlantic horse mackerel (*Trachurus trachurus*) and Atlantic chub mackerel (*Scomber colias*)	Northwest Portuguese coastal waters, Atlantic Ocean	PS, PE	MPs were ingested and caused neurotoxicity and oxidative damage.	Barboza et al. [144]
Discus fish (*Symphysodon aequifasciatus*)	Manacapuru Lake system (Amazon Basin, Brazil), Atlantic Ocean	PS	MP induced oxidative stress in combination with Cd contamination.	Wen et al. [145]
Fathead minnow (*Pimephales promela*)	Laboratory conditions	PS	MP suppresses the immunity in fish.	Greven et al. [149]
Gilthead seabream (*Sparus aurata*) and European sea bass (*D. labrax*)	Local farm (Murcia, Spain), Atlantic Ocean	PVC, PE	MP impacts the fish leukocytes and induce oxidative stress.	Espinosa et al. [139]
Marine medaka (*Oryzias melastigma*)	Laboratory conditions	PS	MP caused damage to reproduction.	Wang et al. [153]
Catfish *(Arius maculatus)*	Songkhla Lake, Thailand, Indian Ocean	Rayon, polyester, polyvinyl alcohol, PE, paint	Accumulation of MP in the stomach.	Pradit et al. [135]
Zebrafish (*D. rerio*) larvae	Laboratory conditions	PS	MP accumulated in the cardiovascular organs.	Veneman et al. [154]
Invertebrate				
Mussels (*Mytilus edulis*)	Port Quinn, Cornwall (UK), Atlantic Ocean	PS	Ingested PS accumulated in the circulatory fluid, and fecal matters contained PS.	Browne et al. [140]
Sea cucumbers (*Holothuroidea* spp.)	Panacea, Florida; Fort Pierce, Florida; and Walpole, Maine (USA), Pacific Ocean	PVC, Nylon	Ingestion of various sizes of PVC and nylon (up to 4 mm), depending on the opening of the tentacles. Poses a threat to primary consumers of sea cucumbers.	Graham and Thompson [146]
Oysters (*Ostrea edulis*)	Queen's University Marine Laboratory, Portaferry (Ireland), Atlantic Ocean	HDPE	Ingestion of HDPE resulted in greater respiration rates in oysters, effecting the mortality rate.	Green [155]

Organisms	Sample location, major ocean sink	Type of MP	Impact	Reference
Copepod (*Calanus helgolandicus*)	English Channel (UK). Atlantic Ocean	PS	PS resulted in decreasing the reproduction rate, but no significant effect on egg production rate, survival rate, and respiration rate.	Cole et al. [114]
Shrimps (*Metapenaeus monoceros, Parapeneopsis stylifera*, and *Penaeus indicus*)	Fishing ground, Arabian Sea, Indian Ocean	PP, PE, polyamide, nylon, polyester, and PET	Microplastics accumulated in the gastrointestinal tract and gut. Shapes of microplastics detected were fiber, pellets, fragments, beads, and films.	Gurjar et al. [156]
Barnacle shrimp (*Amphibalanus amphitrite*) and brine shrimp (*Artemia franciscana*)	Cysts of the species were collected from laboratory from Italy and Belgium, Atlantic Ocean.	PS	MP increase the acetylcholinesterase activity in fish brains, leading to oxidative stress.	Gambardella et al. [157]
Marine copepod (*Tigriopus japonicus*)	Laboratory conditions	PP	MP ingestion and reduction in their fecundity.	Sun et al. [150]
Rotifers (*Brachionus rotundiformis*), Copepods (*Parvocalanus crassirostris*), Shrimp (*Penaeus vannamei*), Scallops (*Chylamys nobilis*)	Center for Collections of Marine Algae at Xiamen University (CCMA, Xiamen, China), Pacific Ocean	PP	MPs were found in the digestive tract.	Ma et al. [129]
Spear shrimp (*Parapenaeopsis hardwickii*), Yellow shrimp (*Metapenaeus brevicornis*)	Songkhla Lake, Thailand, Indian Ocean	Rayon, polyester, polyvinyl alcohol, PE, paint	Accumulation of MP in the stomach.	Pradit et al. [135]
Insects (*Trichoptera, Plecoptera*, and *Coleoptera*)	Vipacco/Vipava River (Friuli Venezia Giulia, northeast Italy, Atlantic Ocean	Polyester	MP accumulation in the invertebrates.	Bertoli et al. [141]
Gammaridae, Asellidae, Tubificidae, and *Chironomidae*	Lowland River (Belgium), Atlantic Ocean.	PE, PP, PVC, others	MP accumulation in the gut.	Pan et al. [158]
38 species of gastropods, bivalves, and crabs	Mudflats and sandy beaches (Hongkong), Pacific Ocean	PET, cellophane, polyamide	0–18 MP per organism was found.	Xu et al. [131]
Chironomids larvae	Lake Jinhu in Chongqing, China, Pacific Ocean	PE	MP lowered the nitrogen removal capability of the larvae.	Huang et al. [130]

Organisms	Sample location, major ocean sink	Type of MP	Impact	Reference
Aquatic larvae caddisfly (*Sericostoma pyrenaicum*)	Perea stream (Spain), Atlantic Ocean	PS	MP were found in the larvae feces, indicating MP ingestions and egestion.	López-Rojo et al. [159]
Marine copepod (*Pseudodiaptomus annandalei*)	Laboratory conditions	PS	MP ingested as well as egested.	Cheng et al. [160]
Sea urchins	Ría de Vigo (Galicia, NW Iberian Peninsula), Atlantic Ocean	PE	MP ingestion detected.	Beiras and Tato [161]
Copepods (*C. helgolandicus*, *Acartia tonsa*) and European lobster (*Homarus gammarus*)	Western Channel Observatory station (UK), Atlantic Ocean	PS, nylon	MP ingestion detected.	Botterell et al. [162]
Other miscellaneous marine biota				
Seabird (red-footed booby, *Sula sula*) and shorebirds (whimbrel, *Numenius phaeopus* and pacific golden plover *Pluvialis fulva*)	Yongxing Island, South China Sea, Western Pacific Ocean	PP-PE copolymer	Birds ingested the microplastics mistaking it for food items. This resulted in accumulation of microplastics in their stomach, esophagus, gastrointestinal tracts, and intestine. Microplastics consisted primarily of thread- shaped and blue-colored pieces.	Zhu et al. [115]
Dolphin (*Delphinus delphis*)	Galicia, Iberian Peninsula, Atlantic Ocean	Not determined	Microplastic accumulated in the stomach of dolphins, including fragments, beads, and fibers.	Hernandez-Gonzalez et al. [163]
Humpback whale (*Megaptera novaeangliae*)	Sandbank between Den Helder and Texel (Netherlands), Atlantic Ocean	PE, PP, PVC, PET, nylon	Various sizes of plastics (1 mm-17 cm) accumulated in the gastrointestinal tract. Shapes detected were sheets, fragments, and threads. Microplastic caused blockage of the intestinal tract, disrupting the digestion process.	Besseling et al. [142]

Organisms	Sample location, major ocean sink	Type of MP	Impact	Reference
Green turtle (*Chelonia mydas*)	Hainan Island (China), North Pacific Ocean	PS, PE	Presence of microplastics in the beach sand resulted in disruption of the nesting ground for turtle and delay in egg hatching,	Zhang et al. [132]
Green algae (*Cladophora* spp.)	Lakes Michigan and Erie (Laurentian Great Lakes), Atlantic Ocean.	PE, PET, Spandex	Cladophora readily sequestered the microplastics from the water. This in return would lead to trophic transfer when these algae will be consumed by other marine organisms.	Peller et al. [164]
Marine diatoms (*Phaeodactylum tricornutum*)	Center for Collections of Marine Algae at Xiamen University (CCMA, Xiamen, China), Pacific Ocean	PP	MP impacts the photosynthesis ability of the algae.	Ma et al. [129]
Algae (*Skeletonema costatum*)	Laboratory conditions	PE, PS, PVC	Microalgae growth decreased with increasing MP concentration.	Zhu et al. [165]
Marine microalgae (*Dunaliella salina*)	Laboratory conditions. Microalgae were procured from Tamil Nadu (India), Indian Ocean.	PS	Low concentration of MP resulted in lowering the toxic impact of co-pollutant (zinc oxide) on microalgae.	Gunasekaran et al. [166]
Marine microalgae (*D. salina*)	Laboratory conditions. Microalgae were procured Library of Marine Samples, Korea Institute of Ocean Science & Technology (KIOST, Geoje, Korea).	PE	In the presence of co-pollutants, MP can remove and leach these pollutants and henceforth enhance the growth of microalgae.	Chae et al. [120]
Walrus (*Odobenus rosmarus*)	Svalbard coastline, Arctic Ocean	PE, PP, polyamide, polyester, acrylic	MP detection in the walrus feces.	Carlsson et al. [167]
Eared Seal (3 species of otariids: *Arctocephalus australis, Arctocephalus phillippii, Otaria byronia*)	Peru and Chile coastlines, Pacific Ocean	PET, nylon	MP detected in seals.	Perez-Venegas et al. [127]

Organisms	Sample location, major ocean sink	Type of MP	Impact	Reference
Beluga whales (*Delphinapterus leucas*)	Hendrickson Island, Northwest Territories (Canada), Pacific and Arctic Ocean	PVC, PP, nylon, polyolefin, PET, polyester	MPs were detected in the gastrointestinal tract.	Moore et al. [168]
Otters (*Lutra lutra*)	West Coast of Norway, Atlantic Ocean	PVC, PS, PET	MPs were detected in the stomach of the otters.	Haave et al. [143]
Fur Seals (*A. australis*)	Chilean Northern Patagonia, Pacific Ocean	Microfibers (Type of MP not determined)	MP detected in the seal's feces.	Perez-Venegas et al. [126]
Harbor seal (*Phoca vitulina vitulina*) and Gray seal (*Halichoerus grypus atlantica*)	Cape Cod, Massachusetts, USA, Atlantic Ocean	Resin, Cellophane, PET, PP	MP detected in the fecal samples.	Hudak and Sette [147]
Harbor porpoises (*Phocoena phocoena*)	Netherlands, Atlantic Ocean	PE, PP, PVC, Polyamide, PET	MP detected in the stomach.	Van Franeker et al. [169]

Table 2.
Impact of microplastic (MP) on various marine biota. Please note that for the laboratory simulated studies, the major ocean sink information has not been included.

commercial fishes, seabass, and mackerel found along the Portugal coast ([113, 144]), discus fish found in the Amazon basin, Brazil [145], sea cucumbers in Florida and Maine (the USA, [146]), and seals in Massachusetts (the USA, [147]).

Lastly, there are several studies conducted under laboratory conditions to understand the impact of microplastics on marine organisms. These include investigating the impact of PS and PE on goldfish [148], the effect of PS on fathead minnow [149], and the effect of PP on marine copepod [150]. Please note that for the laboratory simulated studies, the major ocean sink information has not been included in **Table 2**.

6. Conclusion

The microplastics in the marine environment pose adverse effects on marine organisms, which eventually impact human health. Therefore, for the well-being of humans as well as the conservation of the environment, microplastic pollution is extensively investigated by researchers and scientists around the world. This study summarizes the sources of microplastics (primary and secondary), along with their characterization based on chemical composition, size, and shape. The abiotic and biotic degradation of these microplastics is discussed, showing how various macroplastics (i.e., plastic debris) break down into smaller fragments under the effects of various environmental factors (e.g., temperature, sunlight, and biological agents), chemical damage, and mechanical abrasion. Once formed, the marine habitat serves as the primary sink for the microplastics. The distribution and fate of the microplastics in the marine environment depend on the density, size, shape, and chemical composition of the microplastics, as well as the environmental factors (primarily wind and ocean current velocities). If the density of the microplastic is lower than that of the regional water, the microplastics remain suspended in the gyre (surface waves) and are prone to horizontal distribution because of wind and ocean current velocities. If the density of the microplastic is higher than that of the regional water or its density increases because of biofouling and aggregate formation, it would sink to the bottom of the marine habitat. Once sunk, microplastics can either accumulate in the marine sediment, or they can be redistributed because of bottom water current or bioturbation. Therefore, it is challenging to predict the fate of marine microplastics and requires the attention of researchers to fill the knowledge gap, specifically on the ecotoxicological impact of microplastic on the marine environment. An investigation is needed to study the mechanism of microplastic and chemical pollutant sorption by marine organisms as well as their mode of interaction, evaluate the route of transfer of these contaminants along the food web, and investigate the risk of microplastics on marine organisms as well as human.

Acknowledgements

This work is supported by the Ministry of Science and Technology, Taiwan, under the grant numbers MOST 108-2313-B-002-026 and 109-2313-B-002-049-MY2.

Conflict of interest

The authors declare no conflict of interest.

Author details

Fatima Haque and Chihhao Fan*
Department of Bioenvironmental Systems Engineering, National Taiwan University, Taipei, Taiwan

*Address all correspondence to: chfan@ntu.edu.tw

References

[1] Plastics-the Facts 2019 An Analysis of European Plastics Production, Demand and Waste Data. Belgium: PlasticsEurope; 2019. [Online] Available from: https://plasticseurope.org/wp-content/uploads/2021/10/2019-Plastics-the-facts.pdf. [Accessed: Jul 22, 2022]

[2] Shanmugam V et al. Polymer recycling in additive manufacturing: An opportunity for the circular economy. Materials Circular Economy. 2020;**2**(1):1-11. DOI: 10.1007/S42824-020-00012-0

[3] Jambeck JR et al. Plastic waste inputs from land into the ocean. Science (1979). 2015;**347**(6223):768-771. DOI: 10.1126/SCIENCE.1260352/SUPPL_FILE/JAMBECK.SM.PDF

[4] Evode N, Qamar SA, Bilal M, Barceló D, Iqbal HMN. Plastic waste and its management strategies for environmental sustainability. Case Studies in Chemical and Environmental Engineering. 2021;**4**:100142. DOI: 10.1016/J.CSCEE.2021.100142

[5] OECD. Plastic Pollution Is Growing Relentlessly as Waste Management and Recycling Fall Short, Says OECD. Organisation for Economic Co-operation and Development. 2022. Available from: https://www.oecd.org/environment/plastic-pollution-is-growing-relentlessly-as-waste-management-and-recycling-fall-short.htm. [Accessed: Jul 22, 2022]

[6] J. N. Hahladakis, "Delineating and preventing plastic waste leakage in the marine and terrestrial environment," Environmental Science and Pollution Research, vol. 27, no. 11, pp. 12830-12837, Apr. 2020, doi: 10.1007/S11356-020-08139-Y/FIGURES/1.

[7] Krause JC, von Nordheim H, Bräger S. Marine Nature Conservation in Europe 2006 Proceedings of the Symposium, May 2006. Germany; 2006. [Online] Available from: https://www.researchgate.net/publication/278328811_Marine_Nature_Conservation_in_Europe_2006_Proceedings_of_the_Symposium_May_2006. [Accessed: July 22, 2022]

[8] Barboza LGA, Cózar A, Gimenez BCG, Barros TL, Kershaw PJ, Guilhermino L. Macroplastics pollution in the marine environment. In: World Seas: An Environmental Evaluation, Ecological Issues and Environmental Impacts. Vol. III. Ecological Issues and Environmental Impacts. Academic Press; 2019. pp. 305-328. DOI: 10.1016/B978-0-12-805052-1.00019-X

[9] Golubev S. Macroplastic in seabirds at Mirny, Antarctica. Birds. 2020;**1**(1):13-18. DOI: 10.3390/BIRDS1010003

[10] Prampramote J et al. Association of ocean macroplastic debris with stranded sea turtles in the central gulf of Thailand. Endangered Species Research. 2022;**47**:333-343. DOI: 10.3354/ESR01182

[11] Andrady AL. The plastic in microplastics: A review. Marine Pollution Bulletin. 2017;**119**(1):12-22. DOI: 10.1016/J.MARPOLBUL.2017.01.082

[12] Zhang K et al. Understanding plastic degradation and microplastic formation in the environment: A review. Environmental Pollution. 2021;**274**:116554. DOI: 10.1016/J.ENVPOL.2021.116554

[13] Derraik JGB. The pollution of the marine environment by plastic debris:

A review. Marine Pollution Bulletin. 2002;**44**(9):842-852. DOI: 10.1016/S0025-326X(02)00220-5

[14] Alomar C, Estarellas F, Deudero S. Microplastics in the Mediterranean Sea: Deposition in coastal shallow sediments, spatial variation and preferential grain size. Marine Environmental Research. 2016;**115**:1-10. DOI: 10.1016/J.MARENVRES.2016.01.005

[15] Fendall LS, Sewell MA. Contributing to marine pollution by washing your face: Microplastics in facial cleansers. Marine Pollution Bulletin. 2009;**58**(8):1225-1228. DOI: 10.1016/J.MARPOLBUL.2009.04.025

[16] Julienne F, Delorme N, Lagarde F. From macroplastics to microplastics: Role of water in the fragmentation of polyethylene. Chemosphere. 2019;**236**:124409. DOI: 10.1016/J.CHEMOSPHERE.2019.124409

[17] Woodall LC et al. The deep sea is a major sink for microplastic debris. Royal Society Open Science. 2014;**1**(4). DOI: 10.1098/RSOS.140317

[18] Martin J, Lusher A, Thompson RC, Morley A. The deposition and accumulation of microplastics in marine sediments and bottom water from the Irish continental shelf. Scientific Reports. 2017;**7**(1):1-9. DOI: 10.1038/s41598-017-11079-2

[19] Browne MA, Galloway TS, Thompson RC. Spatial patterns of plastic debris along estuarine shorelines. Environmental Science and Technology. 2010;**44**(9):3404-3409. DOI: 10.1021/ES903784E/SUPPL_FILE/ES903784E_SI_001.PDF

[20] Browne MA et al. Accumulation of microplastic on shorelines woldwide: Sources and sinks. Environmental Science and Technology. 2011;**45**(21):9175-9179. DOI: 10.1021/ES201811S/ASSET/IMAGES/LARGE/ES-2011-01811S_0002.JPEG

[21] Barnes DKA, Galgani F, Thompson RC, Barlaz M. Accumulation and fragmentation of plastic debris in global environments. Philosophical Transactions of the Royal Society B: Biological Sciences. 2009;**364**(1526):1985-1998. DOI: 10.1098/RSTB.2008.0205

[22] Law KL et al. Plastic accumulation in the North Atlantic subtropical gyre. Science (1979). 2010;**329**(5996):1185-1188. DOI: 10.1126/SCIENCE.1192321/SUPPL_FILE/LAW_SOM_REVISION_1.PDF

[23] Reichert J et al. Reef-building corals act as long-term sink for microplastic. Global Change Biology. 2022;**28**(1):33-45. DOI: 10.1111/GCB.15920

[24] United Nations, THE 17 GOALS | Sustainable Development. 2022. Available from: https://sdgs.un.org/goals. [Accessed: July 22, 2022].

[25] Egbeocha CO, Malek S, Emenike CU, Milow P. Feasting on microplastics: Ingestion by and effects on marine organisms. Aquatic Biology. 2018;**27**:93-106. DOI: 10.3354/AB00701

[26] Okeke ES et al. Microplastics in agroecosystems-impacts on ecosystem functions and food chain. Resources, Conservation and Recycling. 2022;**177**:105961. DOI: 10.1016/J.RESCONREC.2021.105961

[27] Mato Y, Isobe T, Takada H, Kanehiro H, Ohtake C, Kaminuma T. Plastic resin pellets as a transport medium for toxic chemicals in the marine environment. Environmental Science and Technology. 2001;**35**(2):318-324. DOI: 10.1021/ES0010498/ASSET/

IMAGES/LARGE/ES0010498F00005.JPEG

[28] Yurtsever M. Glitters as a source of primary microplastics: An approach to environmental responsibility and ethics. Journal of Agricultural and Environmental Ethics. 2019;**32**(3):459-478. DOI: 10.1007/S10806-019-09785-0/FIGURES/1

[29] Gregory MR. Plastic 'scrubbers' in hand cleansers: A further (and minor) source for marine pollution identified. Marine Pollution Bulletin. 1996;**32**(12):867-871. DOI: 10.1016/S0025-326X(96)00047-1

[30] Browne MA, Galloway T, Thompson R. Microplastic—An emerging contaminant of potential concern? Integrated Environmental Assessment and Management. 2007;**3**(4):559-561. DOI: 10.1002/IEAM.5630030412

[31] Bakaraki Turan N, Sari Erkan H, Onkal Engin G. Microplastics in wastewater treatment plants: Occurrence, fate and identification. Process Safety and Environmental Protection. 2021;**146**:77-84. DOI: 10.1016/J.PSEP.2020.08.039

[32] Cole M, Lindeque P, Halsband C, Galloway TS. Microplastics as contaminants in the marine environment: A review. Marine Pollution Bulletin. 2011;**62**(12):2588-2597. DOI: 10.1016/J.MARPOLBUL.2011.09.025

[33] Sun J, Dai X, Wang Q, van Loosdrecht MCM, Ni BJ. Microplastics in wastewater treatment plants: Detection, occurrence and removal. Water Research. 2019;**152**:21-37. DOI: 10.1016/J.WATRES.2018.12.050

[34] Hoellein T, Rojas M, Pink A, Gasior J, Kelly J. Anthropogenic litter in urban freshwater ecosystems: Distribution and microbial interactions. PLoS One. 2014;**9**(6):e98485. DOI: 10.1371/JOURNAL.PONE.0098485

[35] Galvão A, Aleixo M, de Pablo H, Lopes C, Raimundo J. Microplastics in wastewater: Microfiber emissions from common household laundry. Environmental Science and Pollution Research. 2020;**27**(21):26643-26649. DOI: 10.1007/S11356-020-08765-6/FIGURES/3

[36] Jönsson C, Arturin OL, Hanning AC, Landin R, Holmström E, Roos S. Microplastics shedding from textiles—Developing analytical method for measurement of shed material representing release during domestic washing. Sustainability. 2018;**10**(7):2457. DOI: 10.3390/SU10072457

[37] Tiffin L, Hazlehurst A, Sumner M, Taylor M. Reliable quantification of microplastic release from the domestic laundry of textile fabrics. The Journal of the Textile Institute. 2021;**113**(4):558-566. DOI: 10.1080/00405000.2021.1892305

[38] Xue B et al. Underestimated microplastic pollution derived from fishery activities and 'hidden' in deep sediment. Environmental Science and Technology. 2020;**54**(4):2210-2217. DOI: 10.1021/ACS.EST.9B04850/ASSET/IMAGES/LARGE/ES9B04850_0002.JPEG

[39] Jan Kole P, Löhr AJ, van Belleghem FGAJ, Ragas AMJ. Wear and tear of tyres: A stealthy source of microplastics in the environment. International Journal of Environmental Research and Public Health. 2017;**14**(10):1265. DOI: 10.3390/IJERPH14101265

[40] Verschoor AJ, Milieutafel D, and Roex E. Quick Scan and Prioritization

of Microplastic Sources and Emissions. 2014. [Online] Available from: https://www.researchgate.net/publication/277194031 [Accessed: July 22, 2022]

[41] Chen G, Feng Q, Wang J. Mini-review of microplastics in the atmosphere and their risks to humans. Science of the Total Environment. 2020;**703**:135504. DOI: 10.1016/J.SCITOTENV.2019.135504

[42] Zhang Q, Wang R, Shen Y, Zhan L, Xu Z. Characteristics of unorganized emissions of microplastics from road fugitive dust in urban mining bases. Science of the Total Environment. 2022;**827**:154355. DOI: 10.1016/J.SCITOTENV.2022.154355

[43] Zhang J, Wang L, Kannan K. Microplastics in house dust from 12 countries and associated human exposure. Environment International. 2020;**134**:105314. DOI: 10.1016/J.ENVINT.2019.105314

[44] Nematollahi MJ et al. Microplastic occurrence in settled indoor dust in schools. Science of the Total Environment. 2022;**807**:150984. DOI: 10.1016/J.SCITOTENV.2021.150984

[45] Vianello A, Jensen RL, Liu L, Vollertsen J. Simulating human exposure to indoor airborne microplastics using a breathing thermal manikin. Scientific Reports. 2019;**9**(1):1-11. DOI: 10.1038/s41598-019-45054-w

[46] Aragaw TA. Surgical face masks as a potential source for microplastic pollution in the COVID-19 scenario. Marine Pollution Bulletin. 2020;**159**:111517. DOI: 10.1016/J.MARPOLBUL.2020.111517

[47] Han J, He S. Need for assessing the inhalation of micro(nano)plastic debris shed from masks, respirators, and home-made face coverings during the COVID-19 pandemic. Environmental Pollution (Barking, Essex: 1987). 2021;**268**:115728. DOI: 10.1016/J.ENVPOL.2020.115728

[48] Haque F, Fan C. Prospect of microplastic pollution control under the 'new normal' concept beyond COVID-19 pandemic. Journal of Cleaner Production. 2022;**367**:133027. DOI: 10.1016/J.JCLEPRO.2022.133027

[49] Schwarz AE, Ligthart TN, Boukris E, van Harmelen T. Sources, transport, and accumulation of different types of plastic litter in aquatic environments: A review study. Marine Pollution Bulletin. 2019;**143**:92-100. DOI: 10.1016/J.MARPOLBUL.2019.04.029

[50] Aragaw TA, De-la-Torre GE, Teshager AA. Personal protective equipment (PPE) pollution driven by the COVID-19 pandemic along the shoreline of Lake Tana, Bahir Dar, Ethiopia. Science of the Total Environment. 2022;**820**:153261. DOI: 10.1016/J.SCITOTENV.2022.153261

[51] Moore CJ. Synthetic polymers in the marine environment: A rapidly increasing, long-term threat. Environmental Research. 2008;**108**(2):131-139. DOI: 10.1016/J.ENVRES.2008.07.025

[52] de Villiers S. Microfibre pollution hotspots in river sediments adjacent to South africa's coastline. Water SA. 2019;**45**(1):97-102. DOI: 10.4314/WSA.V45I1.11

[53] Zhou Q, Zhang H, Waniek JJ, Luo Y. The distribution and characteristics of microplastics in coastal beaches and mangrove wetlands. Handbook of Environmental Chemistry. 2020;**95**:77-92. DOI: 10.1007/698_2020_459/FIGURES/2

[54] Napper IE, Wright LS, Barrett AC, Parker-Jurd FNF, Thompson RC. Potential microplastic release from the maritime industry: Abrasion of rope. Science of the Total Environment. 2022;**804**:150155. DOI: 10.1016/J.SCITOTENV.2021.150155

[55] Alabi OA, Ologbonjaye K, Awosolu O, Alabi OA, Ologbonjaye KI, Alalade OE. Public and environmental health effects of plastic wastes disposal: A review. Journal of Toxicology and Risk Assessment. 2019;**5**:1-13. DOI: 10.23937/2572-4061.1510021

[56] Gündoğdu S. "Polymer types of microplastic in coastal areas," In Microplastic Pollution. Cham: Springer; 2022. pp. 77-88. DOI: 10.1007/978-3-030-89220-3_4.

[57] Burns EE, Boxall ABA. Microplastics in the aquatic environment: Evidence for or against adverse impacts and major knowledge gaps. Environmental Toxicology and Chemistry. 2018;**37**(11):2776-2796. DOI: 10.1002/ETC.4268

[58] US EPA, Plastic Pellets in the Aquatic Environment: Sources and Recommendations, Duxbury, Massachusetts: Battelle Ocean Sciences; 1992. [Online] Available from: https://citeseerx.ist.psu.edu/viewdoc/download?doi=10.1.1.387.3938&rep=rep1&type=pdf. [Accessed: Jul 22, 2022]

[59] Nuelle MT, Dekiff JH, Remy D, Fries E. A new analytical approach for monitoring microplastics in marine sediments. Environmental Pollution. 2014;**184**:161-169. DOI: 10.1016/J.ENVPOL.2013.07.027

[60] British Plastics Federation. Nylons (Polyamide). London: British Plastics Federation; 2017. Available from: https://www.bpf.co.uk/plastipedia/polymers/Polyamides.aspx. [Accessed: Jul 22, 2022]

[61] Coyle R, Hardiman G, Driscoll KO. Microplastics in the marine environment: A review of their sources, distribution processes, uptake and exchange in ecosystems. Case Studies in Chemical and Environmental Engineering. 2020;**2**:100010. DOI: 10.1016/J.CSCEE.2020.100010

[62] Gu JD. Microbiological deterioration and degradation of synthetic polymeric materials: Recent research advances. International Biodeterioration & Biodegradation. 2003;**52**(2):69-91. DOI: 10.1016/S0964-8305(02)00177-4

[63] Shah AA, Hasan F, Hameed A, Ahmed S. Biological degradation of plastics: A comprehensive review. Biotechnology Advances. 2008;**26**(3):246-265. DOI: 10.1016/J.BIOTECHADV.2007.12.005

[64] Singh B, Sharma N. Mechanistic implications of plastic degradation. Polymer Degradation and Stability. 2008;**93**(3):561-584. DOI: 10.1016/J.POLYMDEGRADSTAB. 2007.11.008

[65] Cooper DA, Corcoran PL. Effects of mechanical and chemical processes on the degradation of plastic beach debris on the island of Kauai, Hawaii. Marine Pollution Bulletin. 2010;**60**(5):650-654. DOI: 10.1016/J.MARPOLBUL.2009.12.026

[66] Porter A, Lyons BP, Galloway TS, Lewis C. Role of marine snows in microplastic fate and bioavailability. Environmental Science and Technology. 2018;**52**(12):7111-7119. DOI: 10.1021/ACS.EST.8B01000/ASSET/IMAGES/LARGE/ES-2018-01000F_0003.JPEG

[67] Tamis JE et al. Environmental risks of car tire microplastic particles and other road runoff pollutants. Microplastics and Nanoplastics.

2021;**1**(1):1-17. DOI: 10.1186/S43591-021-00008-W

[68] Wagner S, Hüffer T, Klöckner P, Wehrhahn M, Hofmann T, Reemtsma T. Tire wear particles in the aquatic environment—A review on generation, analysis, occurrence, fate and effects. Water Research. 2018;**139**:83-100. DOI: 10.1016/J.WATRES.2018.03.051

[69] Torikai A, Takeuchi A, Nagaya S, Fueki K. Photodegradation of polyethylene: Effect of crosslinking on the oxygenated products and mechanical properties. Polymer Photochemistry. 1986;**7**(3):199-211. DOI: 10.1016/0144-2880(86)90027-8

[70] Pirsaheb M, Hossini H, Makhdoumi P. Review of microplastic occurrence and toxicological effects in marine environment: Experimental evidence of inflammation. Process Safety and Environmental Protection. 2020;**142**:1-14. DOI: 10.1016/J.PSEP.2020.05.050

[71] Peterson JD, Vyazovkin S, Wight CA. Kinetics of the thermal and thermo-oxidative degradation of polystyrene, polyethylene and poly(propylene)—peterson—2001—macromolecular chemistry and physics—Wiley Online Library. Macromolecular Chemistry and Physics. 2001 [Online] Available from: https://onlinelibrary.wiley.com/doi/10.1002/1521-3935(20010301)202:6%3C775::AID-MACP775%3E3.0.CO;2-G;**202**(6):775-784 [Accessed: July 22, 2022]

[72] Kamweru PK, Ndiritu FG, Kinyanjui TK, Muthui ZW, Ngumbu RG, Odhiambo PM. Study of temperature and UV wavelength range effects on degradation of photo-irradiated polyethylene films using DMA. Journal of Macromolecular Science. 2011;**50**(7):1338-1349. DOI: 10.1080/00222348.2010.516172

[73] Lee QY, Li H. Photocatalytic degradation of plastic waste: a mini review. Micromachines. 30 Jul 2021;**12**(8):907. DOI: 10.3390/mi12080907

[74] Krupa SV, Manning WJ. Atmospheric ozone: Formation and effects on vegetation. Environmental Pollution. 1988;**50**(1-2):101-137. DOI: 10.1016/0269-7491(88)90187-X

[75] Hocker S, Rhudy AK, Ginsburg G, Kranbuehl DE. Polyamide hydrolysis accelerated by small weak organic acids. Polymer (Guildf). 2014;**55**(20):5057-5064. DOI: 10.1016/J.POLYMER.2014.08.010

[76] Gewert B, Plassmann MM, Macleod M. Pathways for degradation of plastic polymers floating in the marine environment. Environmental Science: Processes & Impacts. 2015;**17**(9):1513-1521. DOI: 10.1039/C5EM00207A

[77] Chandra P, Enespa SD, Singh DP. Microplastic degradation by bacteria in aquatic ecosystem. *Microorganisms for Sustainable Environment and Health*. 2020:431-467. DOI: 10.1016/B978-0-12-819001-2.00022-X

[78] Nguyen V, Karunakaran E, Collins G, Biggs CA. Physicochemical analysis of initial adhesion and biofilm formation of Methanosarcina barkeri on polymer support material. Colloids and Surfaces B: Biointerfaces. 2016;**143**:518-525. DOI: 10.1016/J.COLSURFB.2016.03.042

[79] Reisser J et al. Millimeter-sized marine plastics: A new pelagic habitat for microorganisms and invertebrates. PLoS One. 2014;**9**(6):e100289. DOI: 10.1371/JOURNAL.PONE.0100289

[80] Santo M, Weitsman R, Sivan A. The role of the copper-binding enzyme—Laccase—In the biodegradation of polyethylene by the actinomycete Rhodococcus ruber. International Biodeterioration & Biodegradation. 2013;**84**:204-210. DOI: 10.1016/J.IBIOD.2012.03.001

[81] Denaro R et al. Marine hydrocarbon-degrading bacteria breakdown poly(ethylene terephthalate) (PET). Science of the Total Environment. 2020;**749**:141608. DOI: 10.1016/J.SCITOTENV.2020.141608

[82] Mor R, Sivan A. Biofilm formation and partial biodegradation of polystyrene by the actinomycete Rhodococcus ruber: Biodegradation of polystyrene. Biodegradation. 2008;**19**(6):851-858. DOI: 10.1007/S10532-008-9188-0/FIGURES/8

[83] Zeghal E, Vaksmaa A, Vielfaure H, Boekhout T, Niemann H. The potential role of marine Fungi in plastic degradation—A review. Frontiers in Marine Science. 2021;**8**:1783. DOI: 10.3389/FMARS.2021.738877/BIBTEX

[84] Chia WY, Ying Tang DY, Khoo KS, Kay Lup AN, Chew KW. Nature's fight against plastic pollution: Algae for plastic biodegradation and bioplastics production. Environmental Science and Ecotechnology. 2020;**4**:100065. DOI: 10.1016/J.ESE.2020.100065

[85] Smith IL, Stanton T, Law A. Plastic habitats: Algal biofilms on photic and aphotic plastics. Journal of Hazardous Materials Letters. 2021;**2**:100038. DOI: 10.1016/J.HAZL.2021.100038

[86] Ali SS et al. Plastic wastes biodegradation: Mechanisms, challenges and future prospects. Science of the Total Environment. 2021;**780**:146590. DOI: 10.1016/J.SCITOTENV.2021.146590

[87] Auta HS, Emenike CU, Fauziah SH. Distribution and importance of microplastics in the marine environment: A review of the sources, fate, effects, and potential solutions. Environment International. 2017;**102**:165-176. DOI: 10.1016/J.ENVINT.2017.02.013

[88] Sutton R, Mason SA, Stanek SK, Willis-Norton E, Wren IF, Box C. Microplastic contamination in the San Francisco Bay, California, USA. Marine Pollution Bulletin. 2016;**109**(1):230-235. DOI: 10.1016/J.MARPOLBUL.2016.05.077

[89] Kukulka T, Proskurowski G, Morét-Ferguson S, Meyer DW, Law KL. The effect of wind mixing on the vertical distribution of buoyant plastic debris. *Geophysical Research Letters*. 2012;**39**(7). DOI: 10.1029/2012GL051116

[90] Guzzetti E, Sureda A, Tejada S, Faggio C. Microplastic in marine organism: Environmental and toxicological effects. Environmental Toxicology and Pharmacology. 2018;**64**:164-171. DOI: 10.1016/J.ETAP.2018.10.009

[91] Obbard RW, Sadri S, Wong YQ, Khitun AA, Baker I, Thompson RC. Global warming releases microplastic legacy frozen in Arctic Sea ice. Earth's Future. 2014;**2**(6):315-320. DOI: 10.1002/2014EF000240

[92] Eriksen M et al. Microplastic pollution in the surface waters of the Laurentian Great Lakes. Marine Pollution Bulletin. 2013;**77**(1-2, 182):177. DOI: 10.1016/J.MARPOLBUL.2013.10.007

[93] Goldstein MC, Titmus AJ, Ford M. Scales of spatial heterogeneity of plastic marine debris in the Northeast Pacific Ocean. PLoS One. 2013;**8**(11):e80020. DOI: 10.1371/JOURNAL.PONE.0080020

[94] Yang H, Chen G, Wang J. Microplastics in the marine environment: Sources, fates, impacts and microbial degradation. Toxics. 2021;**9**(2):41. DOI: 10.3390/TOXICS9020041

[95] Gouin T, Roche N, Lohmann R, Hodges G. A thermodynamic approach for assessing the environmental exposure of chemicals absorbed to microplastic. Environmental Science and Technology. 2011;**45**(4):1466-1472. DOI: 10.1021/ES1032025/SUPPL_FILE/ES1032025_SI_001.PDF

[96] Xia W, Rao Q, Deng X, Chen J, Xie P. Rainfall is a significant environmental factor of microplastic pollution in inland waters. Science of the Total Environment. 2020;**732**:139065. DOI: 10.1016/J.SCITOTENV.2020.139065

[97] Bullard JE, Ockelford A, O'Brien P, McKenna Neuman C. Preferential transport of microplastics by wind. Atmospheric Environment. 2021;**245**:118038. DOI: 10.1016/J.ATMOSENV.2020.118038

[98] Liu K et al. Accumulation of microplastics in a downstream area of a semi-enclosed bay: Implications of input from coastal currents. Science of the Total Environment. 2021;**791**:148280. DOI: 10.1016/J.SCITOTENV.2021.148280

[99] Rezania S et al. Microplastics pollution in different aquatic environments and biota: A review of recent studies. Marine Pollution Bulletin. 2018;**133**:191-208. DOI: 10.1016/J.MARPOLBUL.2018.05.022

[100] Cózar A et al. Plastic debris in the open ocean. Proceedings of the National Academy of Sciences of the United States of America. 2014;**111**(28):10239-10244. DOI: 10.1073/PNAS.1314705111/-/DCSUPPLEMENTAL

[101] Welden NAC, Lusher AL. Impacts of changing ocean circulation on the distribution of marine microplastic litter. Integrated Environmental Assessment and Management. 2017;**13**(3):483-487. DOI: 10.1002/IEAM.1911

[102] Kooi M, van Nes EH, Scheffer M, Koelmans AA. Ups and downs in the ocean: Effects of biofouling on vertical transport of microplastics. Environmental Science and Technology. 2017;**51**(14):7963-7971. DOI: 10.1021/ACS.EST.6B04702/ASSET/IMAGES/LARGE/ES-2016-047026_0003.JPEG

[103] Besseling E, Quik JTK, Sun M, Koelmans AA. Fate of nano- and microplastic in freshwater systems: A modeling study. Environmental Pollution. 2017;**220**:540-548. DOI: 10.1016/J.ENVPOL.2016.10.001

[104] Kaiser D, Kowalski N, Waniek JJ. Effects of biofouling on the sinking behavior of microplastics. Environmental Research Letters. 2017;**12**(12):124003. DOI: 10.1088/1748-9326/AA8E8B

[105] Zhao S, Danley M, Ward JE, Li D, Mincer TJ. An approach for extraction, characterization and quantitation of microplastic in natural marine snow using Raman microscopy. Analytical Methods. 2017;**9**(9):1470-1478. DOI: 10.1039/C6AY02302A

[106] Alimi OS, Farner Budarz J, Hernandez LM, Tufenkji N. Microplastics and Nanoplastics in aquatic environments: Aggregation, deposition, and enhanced contaminant transport. Environmental Science and Technology. 2018;**52**(4):1704-1724. DOI: 10.1021/ACS.EST.7B05559/ASSET/IMAGES/LARGE/ES-2017-05559J_0004.JPEG

[107] Abed RMM, Muthukrishnan T, Al Khaburi M, Al-Senafi F,

Munam A, Mahmoud H. Degradability and biofouling of oxo-biodegradable polyethylene in the planktonic and benthic zones of the Arabian Gulf. Marine Pollution Bulletin. 2020;**150**:110639. DOI: 10.1016/j.marpolbul.2019.110639

[108] Leiser R, Wu GM, Neu TR, Wendt-Potthoff K. Biofouling, metal sorption and aggregation are related to sinking of microplastics in a stratified reservoir. Water Research. 2020;**176**:115748. DOI: 10.1016/J.WATRES.2020.115748

[109] Lobelle D, Cunliffe M. Early microbial biofilm formation on marine plastic debris. Marine Pollution Bulletin. 2011;**62**(1):197-200. DOI: 10.1016/J.MARPOLBUL.2010.10.013

[110] Artham T, Sudhakar M, Venkatesan R, Madhavan Nair C, Murty KVGK, Doble M. Biofouling and stability of synthetic polymers in sea water. International Biodeterioration & Biodegradation. 2009;**63**(7):884-890. DOI: 10.1016/J.IBIOD.2009.03.003

[111] Näkki P, Setälä O, Lehtiniemi M. Bioturbation transports secondary microplastics to deeper layers in soft marine sediments of the northern Baltic Sea. Marine Pollution Bulletin. 2017;**119**(1):255-261. DOI: 10.1016/J.MARPOLBUL.2017.03.065

[112] Thompson RC, Moore CJ, Saal FSV, Swan SH. Plastics, the environment and human health: Current consensus and future trends. Philosophical Transactions of the Royal Society B: Biological Sciences. 2009;**364**(1526):2153-2166. DOI: 10.1098/RSTB.2009.0053

[113] Neves D, Sobral P, Ferreira JL, Pereira T. Ingestion of microplastics by commercial fish off the Portuguese coast. Marine Pollution Bulletin. 2015;**101**(1):119-126. DOI: 10.1016/J.MARPOLBUL.2015.11.008

[114] Cole M, Lindeque P, Fileman E, Halsband C, Galloway TS. The impact of polystyrene microplastics on feeding, function and fecundity in the marine copepod Calanus helgolandicus. Environmental Science and Technology. 2015;**49**(2):1130-1137. DOI: 10.1021/ES504525U/ASSET/IMAGES/LARGE/ES-2014-04525U_0005.JPEG

[115] Zhu C et al. Plastic debris in marine birds from an island located in the South China Sea. Marine Pollution Bulletin. 2019;**149**:110566. DOI: 10.1016/J.MARPOLBUL.2019.110566

[116] Sussarellu R et al. Oyster reproduction is affected by exposure to polystyrene microplastics. Proceedings of the National Academy of Sciences. 2016;**113**(9):2430-2435. DOI: 10.1073/PNAS.1519019113

[117] van Raamsdonk LWD et al. Current insights into monitoring, bioaccumulation, and potential health effects of microplastics present in the food chain. Foods 2020. 2020;**9**(1):72. DOI: 10.3390/FOODS9010072

[118] Kang J, Zhou L, Duan X, Sun H, Ao Z, Wang S. Degradation of cosmetic microplastics via functionalized carbon nanosprings. Matter. 2019;**1**(3):745-758. DOI: 10.1016/J.MATT.2019.06.004

[119] Gola D et al. The impact of microplastics on marine environment: A review. Environmental Nanotechnology, Monitoring & Management. 2021;**16**:100552. DOI: 10.1016/J.ENMM.2021.100552

[120] Chae Y, Kim D, An YJ. Effects of micro-sized polyethylene spheres on the marine microalga *Dunaliella salina*: Focusing on the algal cell to plastic

particle size ratio. Aquatic Toxicology. 2019;**216**:105296. DOI: 10.1016/J. AQUATOX.2019.105296

[121] Lei L et al. Microplastic particles cause intestinal damage and other adverse effects in zebrafish *Danio rerio* and nematode *Caenorhabditis elegans*. Science of the Total Environment. 2018;**619-620**:1-8. DOI: 10.1016/J. SCITOTENV.2017.11.103

[122] Wang J et al. Microplastics in the surface sediments from the Beijiang River littoral zone: Composition, abundance, surface textures and interaction with heavy metals. Chemosphere. 2017;**171**:248-258. DOI: 10.1016/J. CHEMOSPHERE.2016.12.074

[123] Luís LG, Ferreira P, Fonte E, Oliveira M, Guilhermino L. Does the presence of microplastics influence the acute toxicity of chromium(VI) to early juveniles of the common goby (Pomatoschistus microps)? A study with juveniles from two wild estuarine populations. Aquatic Toxicology. 2015;**164**:163-174. DOI: 10.1016/J. AQUATOX.2015.04.018

[124] Guo X, Wang J. The chemical behaviors of microplastics in marine environment: A review. Marine Pollution Bulletin. 2019;**142**:1-14. DOI: 10.1016/J. MARPOLBUL.2019.03.019

[125] Rochman CM, Kurobe T, Flores I, Teh SJ. Early warning signs of endocrine disruption in adult fish from the ingestion of polyethylene with and without sorbed chemical pollutants from the marine environment. Science of the Total Environment. 2014;**493**:656-661. DOI: 10.1016/J.SCITOTENV.2014.06.051

[126] Perez-Venegas DJ et al. First detection of plastic microfibers in a wild population of south American fur seals (*Arctocephalus australis*) in the Chilean northern Patagonia. Marine Pollution Bulletin. 2018;**136**:50-54. DOI: 10.1016/J. MARPOLBUL.2018.08.065

[127] Perez-Venegas DJ et al. Monitoring the occurrence of microplastic ingestion in otariids along the Peruvian and Chilean coasts. Marine Pollution Bulletin. 2020;**153**:110966. DOI: 10.1016/J. MARPOLBUL.2020.110966

[128] Lu Y et al. Uptake and accumulation of polystyrene microplastics in zebrafish (*Danio rerio*) and toxic effects in liver. Environmental Science and Technology. 2016;**50**(7):4054-4060. DOI: 10.1021/ ACS.EST.6B00183/ASSET/IMAGES/ LARGE/ES-2016-00183C_0007.JPEG

[129] Ma J et al. Face masks as a source of nanoplastics and microplastics in the environment: Quantification, characterization, and potential for bioaccumulation. Environmental Pollution. 2021;**288**:117748. DOI: 10.1016/J.ENVPOL.2021.117748

[130] Huang Y et al. Effect of microplastics on ecosystem functioning: Microbial nitrogen removal mediated by benthic invertebrates. Science of the Total Environment. 2021;**754**:142133. DOI: 10.1016/J.SCITOTENV.2020.142133

[131] Xu X, Wong CY, Tam NFY, Lo HS, Cheung SG. Microplastics in invertebrates on soft shores in Hong Kong: Influence of habitat, taxa and feeding mode. Science of the Total Environment. 2020;**715**:136999. DOI: 10.1016/J.SCITOTENV.2020.136999

[132] Zhang T et al. The microplastic pollution in beaches that served as historical nesting grounds for green turtles on Hainan Island, China. Marine Pollution Bulletin. 2021;**173**:113069. DOI: 10.1016/J. MARPOLBUL.2021.113069

[133] Population, total—China. Available from: https://data.worldbank.org/indicator/SP.POP.TOTL?locations=CN,%20accessed%20 2022/01/30 [Accessed: Aug 9, 2022]

[134] Plastic Pollution—Our World in Data. 2018. Available from: https://ourworldindata.org/plastic-pollution [Accessed: Aug 9, 2022]

[135] Pradit S, Noppradit BP, Goh K, Sornplang M, Ong C, Towatana P. Occurrence of microplastics and trace metals in fish and shrimp from Songkhla lake, Thailand during the Covid-19 pandemic. Applied Ecology and Environmental Research. 2021;**19**(2):1085-1106. DOI: 10.15666/aeer/1902_10851106

[136] Lusher AL, McHugh M, Thompson RC. Occurrence of microplastics in the gastrointestinal tract of pelagic and demersal fish from the English Channel. Marine Pollution Bulletin. 2013;**67**(1-2):94-99. DOI: 10.1016/J.MARPOLBUL.2012.11.028

[137] Mazurais D et al. Evaluation of the impact of polyethylene microbeads ingestion in European sea bass (*Dicentrarchus labrax*) larvae. Marine Environmental Research. 2015;**112**:78-85. DOI: 10.1016/J.MARENVRES.2015.09.009

[138] Collard F et al. Morphology of the filtration apparatus of three planktivorous fishes and relation with ingested anthropogenic particles. Marine Pollution Bulletin. 2017;**116**(1-2):182-191. DOI: 10.1016/J.MARPOLBUL.2016.12.067

[139] Espinosa C, García Beltrán JM, Esteban MA, Cuesta A. In vitro effects of virgin microplastics on fish head-kidney leucocyte activities. Environmental Pollution. 2018;**235**:30-38. DOI: 10.1016/J.ENVPOL.2017.12.054

[140] Browne MA, Dissanayake A, Galloway TS, Lowe DM, Thompson RC. Ingested microscopic plastic translocates to the circulatory system of the mussel, *Mytilus edulis* (L.). Environmental Science and Technology. 2008;**42**(13):5026-5031. DOI: 10.1021/ES800249A/SUPPL_FILE/ES800249A-FILE002.PDF

[141] Bertoli M et al. Microplastics accumulation in functional feeding guilds and functional habit groups of freshwater macrobenthic invertebrates: Novel insights in a riverine ecosystem. Science of the Total Environment. 2022;**804**:150207. DOI: 10.1016/J.SCITOTENV.2021.150207

[142] Besseling E et al. Microplastic in a macro filter feeder: Humpback whale *Megaptera novaeangliae*. Marine Pollution Bulletin. 2015;**95**(1):248-252. DOI: 10.1016/J.MARPOLBUL.2015.04.007

[143] Haave M, Gomiero A, Schönheit J, Nilsen H, Olsen AB. Documentation of microplastics in tissues of wild coastal animals. Frontiers in Environmental Science. 2021;**9**:31. DOI: 10.3389/FENVS.2021.575058/BIBTEX

[144] Barboza LGA et al. Microplastics in wild fish from north East Atlantic Ocean and its potential for causing neurotoxic effects, lipid oxidative damage, and human health risks associated with ingestion exposure. Science of the Total Environment. 2020;**717**:134625. DOI: 10.1016/J.SCITOTENV.2019.134625

[145] Wen B et al. Single and combined effects of microplastics and cadmium on the cadmium accumulation, antioxidant defence and innate immunity of the discus fish (*Symphysodon*

aequifasciatus). Environmental Pollution. 2018;**243**:462-471. DOI: 10.1016/J.ENVPOL.2018.09.029

[146] Graham ER, Thompson JT. Deposit- and suspension-feeding sea cucumbers (Echinodermata) ingest plastic fragments. Journal of Experimental Marine Biology and Ecology. 2009;**368**(1):22-29. DOI: 10.1016/J.JEMBE.2008.09.007

[147] Hudak CA, Sette L. Opportunistic detection of anthropogenic micro debris in harbor seal (*Phoca vitulina vitulina*) and gray seal (Halichoerus grypus atlantica) fecal samples from haul-outs in southeastern Massachusetts, USA. Marine Pollution Bulletin. 2019;**145**:390-395. DOI: 10.1016/J.MARPOLBUL.2019.06.020

[148] Jabeen K et al. Effects of virgin microplastics on goldfish (Carassius auratus). Chemosphere. 2018;**213**:323-332. DOI: 10.1016/J.CHEMOSPHERE.2018.09.031

[149] Greven AC et al. Polycarbonate and polystyrene nanoplastic particles act as stressors to the innate immune system of fathead minnow (*Pimephales promelas*). Environmental Toxicology and Chemistry. 2016;**35**(12):3093-3100. DOI: 10.1002/ETC.3501

[150] Sun J et al. Release of microplastics from discarded surgical masks and their adverse impacts on the marine copepod Tigriopus japonicus. Environmental Science and Technology Letters. 2021;**8**(12):1065-1070. DOI: 10.1021/ACS.ESTLETT.1C00748/ASSET/IMAGES/LARGE/EZ1C00748_0002.JPEG

[151] Romano N, Ashikin M, Teh JC, Syukri F, Karami A. Effects of pristine polyvinyl chloride fragments on whole body histology and protease activity in silver barb Barbodes gonionotus fry. Environmental Pollution. 2018;**237**:1106-1111. DOI: 10.1016/J.ENVPOL.2017.11.040

[152] Pedà C et al. Intestinal alterations in European sea bass Dicentrarchus labrax (Linnaeus, 1758) exposed to microplastics: Preliminary results. Environmental Pollution. 2016;**212**:251-256. DOI: 10.1016/J.ENVPOL.2016.01.083

[153] Wang J et al. Polystyrene microplastics cause tissue damages, sex-specific reproductive disruption and transgenerational effects in marine medaka (*Oryzias melastigma*). Environmental Pollution. 2019;**254**:113024. DOI: 10.1016/J.ENVPOL.2019.113024

[154] Veneman WJ, Spaink HP, Brun NR, Bosker T, Vijver MG. Pathway analysis of systemic transcriptome responses to injected polystyrene particles in zebrafish larvae. Aquatic Toxicology. 2017;**190**:112-120. DOI: 10.1016/J.AQUATOX.2017.06.014

[155] Green DS. Effects of microplastics on European flat oysters, Ostrea edulis and their associated benthic communities. Environmental Pollution. 2016;**216**:95-103. DOI: 10.1016/J.ENVPOL.2016.05.043

[156] Gurjar UR et al. Microplastics in shrimps: A study from the trawling grounds of north eastern part of Arabian Sea. Environmental Science and Pollution Research. 2021;**28**(35):48494-48504. DOI: 10.1007/S11356-021-14121-Z/FIGURES/7

[157] Gambardella C et al. Effects of polystyrene microbeads in marine planktonic crustaceans. Ecotoxicology and Environmental

Safety. 2017;**145**:250-257. DOI: 10.1016/J.ECOENV.2017.07.036

[158] Pan CG et al. Automated μfTIR imaging demonstrates taxon-specific and selective uptake of microplastic by freshwater invertebrates. Environmental Science and Technology. 2021;**55**(14):9916-9925. DOI: 10.1021/ACS.EST.1C03119/ASSET/IMAGES/LARGE/ES1C03119_0004.JPEG

[159] López-Rojo N, Pérez J, Alonso A, Correa-Araneda F, Boyero L. Microplastics have lethal and sublethal effects on stream invertebrates and affect stream ecosystem functioning. Environmental Pollution. 2020;**259**:113898. DOI: 10.1016/J.ENVPOL.2019.113898

[160] Cheng Y, Wang J, Yi X, Li L, Liu X, Ru S. Low microalgae availability increases the ingestion rates and potential effects of microplastics on marine copepod Pseudodiaptomus annandalei. Marine Pollution Bulletin. 2020;**152**:110919. DOI: 10.1016/J.MARPOLBUL.2020.110919

[161] Beiras R, Tato T. Microplastics do not increase toxicity of a hydrophobic organic chemical to marine plankton. Marine Pollution Bulletin. 2019;**138**:58-62. DOI: 10.1016/J.MARPOLBUL.2018.11.029

[162] Botterell ZLR et al. Bioavailability of microplastics to marine zooplankton: Effect of shape and Infochemicals. Environmental Science and Technology. 2020;**54**(19):12024-12033. DOI: 10.1021/ACS.EST.0C02715/ASSET/IMAGES/LARGE/ES0C02715_0005.JPEG

[163] Hernandez-Gonzalez A, Saavedra C, Gago J, Covelo P, Santos MB, Pierce GJ. Microplastics in the stomach contents of common dolphin (*Delphinus delphis*) stranded on the Galician coasts (NW Spain, 2005-2010). Marine Pollution Bulletin. 2018;**137**:526-532. DOI: 10.1016/J.MARPOLBUL.2018.10.026

[164] Peller J et al. Sequestration of microfibers and other microplastics by green algae, Cladophora, in the US Great Lakes. Environmental Pollution. 2021;**276**:116695. DOI: 10.1016/J.ENVPOL.2021.116695

[165] Zhu Z-L, Wang S-C, Zhao F-F, Wang S-G, Liu F-F, Liu G-Z. Joint toxicity of microplastics with triclosan to marine microalgae *Skeletonema costatum*. *Environmental Pollution*. 2019;**246**:509-517. DOI: 10.1016/J.ENVPOL.2018.12.044

[166] Gunasekaran D, Chandrasekaran N, Jenkins D, Mukherjee A. Plain polystyrene microplastics reduce the toxic effects of ZnO particles on marine microalgae Dunaliella salina. Journal of Environmental Chemical Engineering. 2020;**8**(5):104250. DOI: 10.1016/J.JECE.2020.104250

[167] Carlsson P, Singdahl-Larsen C, Lusher AL. Understanding the occurrence and fate of microplastics in coastal Arctic ecosystems: The case of surface waters, sediments and walrus (*Odobenus rosmarus*). Science of the Total Environment. 2021;**792**:148308. DOI: 10.1016/J.SCITOTENV.2021.148308

[168] Moore RC et al. Microplastics in beluga whales (*Delphinapterus leucas*) from the eastern Beaufort Sea. Marine Pollution Bulletin. 2020;**150**:110723. DOI: 10.1016/J.MARPOLBUL.2019.110723

[169] van Franeker JA et al. Plastic ingestion by harbour porpoises Phocoena phocoena in the Netherlands: Establishing a standardised method. Ambio. 2018;**47**(4):387-397. DOI: 10.1007/S13280-017-1002-Y/FIGURES/1

Section 4

Mercury Pollution

Chapter 7

Perspective Chapter: The Toxic Silver (Hg)

Ahmed A. Abdelhafez, Abdel Aziz Tantawy, Mohamed H.H. Abbas, Shawky M. Metwally, Amera Sh. Metwally, Aya Sh. Metwally, Rasha R.M. Mansour, Sedky H. Hassan, Hassan H. Abbas, Ihab M. Farid, Nermeen N. Nasralla, Ahmed S.H. Soliman, Mohammed E. Younis, Ghada S.A. Sayed, Mahfouz Z. Ahmed, Ehdaa Alaa Mohamed Abed, Ahmed Farouk Al-Hossainy, Heidi Ahmed Ali Abouzeid, Mahdy H. Hamed, Mahmoud I. El-Kelawy, Gamal Hassan Kamel, Hussein Ferweez and Ahmed M. Diab

Abstract

In the late 1950s, residents of a Japanese fishing village known as "Minamata" began falling ill and dying at an alarming rate. The Japanese authorities stated that methyl-mercury-rich seafood and shellfish caused the sickness. Burning fossil fuels represent ≈52.7% of Hg emissions. The majorities of mercury's compounds are volatile and thus travel hundreds of miles with wind before being deposited on the earth's surface. High acidity and dissolved organic carbon increase Hg-mobility in soil to enter the food chain. Additionally, Hg is taken up by areal plant parts via gas exchange. Mercury has no identified role in plants while exhibiting high affinity to form complexes with soft ligands such as sulfur and this consequently inactivates amino acids and sulfur-containing antioxidants. Long-term human exposure to Hg leads to neurotoxicity in children and adults, immunological, cardiac, and motor reproductive and genetic disorders. Accordingly, remediating contaminated soils has become an obligation. Mercury, like other potentially toxic elements, is not biodegradable, and therefore, its remediation should encompass either removal of Hg from soils or even its immobilization. This chapter discusses Hg's chemical behavior, sources, health dangers, and soil remediation methods to lower Hg levels.

Keywords: Hg sources, soil contamination, Hg poisoning, chemical behavior, remediation techniques

1. Introduction

In the late 1950s, people and animals of a Japanese fishing village, called Minamata fell ill one after another and suffered from the same symptoms then died [1, 2]. The Japanese government named this strange disease "Minamat" and announced officially that its cause was the consumption of fish and shellfish that contained high levels of methyl-mercury due to mixing the bay water therein with industrial wastes coming from a chemical factory [3]. Since then, the scientists and government administrators tried hard to avoid a repeat of such type of incidents [4]. Mercury (Hg) is one of the 10 leading worldwide chemicals of concern [5, 6]. This element is denoted by the symbol "Hg" with atomic number equal to 80. Its melting point is −38.83°C while the boiling point is 356.73°C. In the periodic table, only Hg is a metal found in a liquid state at standard temperature and pressure [6]. Generally, mercury level increases progressively in the environment due to anthropogenic activities [7]. All mercury forms that are released during mining and other industries will eventually wind up in soils or surface water representing a potential threat to human health, and the surrounding ecosystem. Every living organism, in the environment (humans, animals, and plants to the smaller ones such as bacteria), is vulnerable to the effects of mercury poisoning. As a result, countries stand to lose millions of dollars in earning potential every year due to mercury contamination.

2. Forms of mercury in the environment

Mercury (Hg) is recognized by its characteristic shiny silver appearance. In the environment, it occurs in different inorganic (mercurous ion (Hg^{+1}) and mercuric ion (Hg^{+2}), and organic forms (methyl mercury (CH_3Hg^+), ethyl mercury ($C_2H_5Hg^+$), and phenyl mercury ($C_6H_5Hg^+$). The elemental mercury is the purest form [8, 9] and, at the same time, exhibits the least toxicity [10]. Under anoxic and suboxic conditions, inorganic Hg can be transformed into CH_3Hg^+ through sulfate-reducing and iron-reducing bacteria [6].

2.1 Sources of mercury

Mercury is introduced into the environment *via* natural and anthropogenic pathways. The first route results from volcanoes, weathering of rocks, forest fires, and soils [11]. The second one represents one-third of its content in nature which is related to anthropogenic activities coming from industrial processes [12] such as the burning of fossil fuels which represent up to 52.7% of Hg emissions [13], gold mining, cement production [9] and combustion of fossil fuel and agricultural additives that increases Hg levels in soil, for example, municipal [6], sludge, fertilizers, lime, and manures [10].

Long-term mining and smelting activities could bring considerable amount of Hg to the surroundings [14]. The majority of mercury's compounds are very volatile and thus travel hundreds of miles with the wind before being deposited on the earth's surface, hence contaminating the surrounding areas [15]. Generally, mercury in air can be carried by rain and eroded soils and run off to the surface waters and no one becomes safe [13]. In a study by Rodríguez Martín [16], most emitted Hg results from power plants that burn coal to create electricity; they account for about 42% of all manmade mercury emissions.

The main countries, contributing to the majority of emitted mercury (kg $year^{-1}$) according to the United Nations Environment Program report in 2018, are: China (572,195), India (205862), Indonesia (156766), Brazil (71470), and Russia (60,949 tonnes), representing an average percentage of 25.73, 9.25, 7.05, 3.21, and 2.74% of total emitted mercury, respectively [17].

3. Mercury in medicine

Mercury is used in dental amalgam fillings to increase its strength and longevity; yet it plays a negative role in increasing human toxicity [18, 19]. Although this type of pollution is going down as the number of medical waste incinerators is reduced, the health community is concerned about patients and other vulnerable groups who are exposed to in healthcare products [20]. Thiomersal is an organic compound used as a preservative in vaccines, and Merbromin (Mercurochrome) is a topical antiseptic used for minor cuts and scrapes and is still in use in some countries [21]. Mercury compounds are found in some over-the-counter drugs, including topical antiseptics, stimulant laxatives, diaper-rash ointment, eye drops, and nasal sprays [22]. The skin-whitening cosmetic products can also be a source of Hg pollution [23]. Even some traditional medicines in China as Siddha and Ayurveda may contain mercury that causes chronic poisoning [24]. Overall, air, water, food, cosmetics, and even vaccines are potential sources of human pollution with Hg [25]. Mercury and most of its compounds are extremely toxic and must be handled with care.

4. Mercury in food chain

Mercury enters the food chain *via* various pathways. Chloralkali industry is one of the European users that pollute Europe's aquatic environments with tones of mercury [20]. In aquatic environments, inorganic mercury biotransforms into methyl mercury, which makes mercury biomagnify in food chains [13]. High acidity and high concentration of dissolved organic carbon in the water enhance the mobility of mercury that enters the food chain [26, 27]. People, who eat a lot of fresh or marine food, have the high risk of mercury intake [28].

On the other hand, this pollutant (Hg) is highly mobile in soils; and can be absorbed easily by plant roots [12]; yet, in the presence of organic additives, the mobility of inorganic and organic forms of Hg could be diminished considerably, forming low mobile complexes [29]. Also, Hg in the atmosphere is taken up in substantial amounts by areal plant parts *via* gas exchange [30], accumulates in edible plant parts [8], and hence enters the food chain [6, 18, 19]. Additionally, sewage irrigation practices account for further soil contamination with Hg [31]. Anyhow, this contaminant has no identified biological role in plants [8]; nevertheless, it exhibits high affinity to forms complexes with soft ligands such as sulfur in the form of insoluble and stable compounds [32]. This in turn inactivates numerous enzymatic reactions, amino acids, and sulfur-containing antioxidants [33].

5. Mercury in sewage effluents

Municipal sewage has been noted as a significant environmental mercury (Hg) source. Mercury in the effluents of waste water treatment plants and mercury-based fungicides increase the discharge of mercury to the aquatic environment [34, 35]. Consumption of Hg-containing foods [12, 36] and exposure to common items like batteries [29] would increase the risks of mercury being excreted and flushed away in the city's sewage system. The mercury released by hospitals, dentist clinics, and other service facilities is a major source of Hg in sewage [37]. A total of 30 tons of total Hg (organic and inorganic) was loaded into sludge in China in 2019, accounting for around 3.6% of the total anthropogenic Hg release (including direct and secondary anthropogenic releases). It is worth noting that sludge treatment methods such as incineration, carbonization, and sludge/brick/cement manufacture pose the greatest threat to atmospheric Hg pollution [27]. Therefore, attention should be paid regarding Hg pollution of sewage effluents and standard regulations should be formulated in order to prevent the environment and human health.

6. Health risks associated with mercury hazards

The excessive population growth, Industrial Revolution, and unmanaged development led to negative impacts on the surrounding environment [34, 35].

Consumption of food high in its content of mercury is the main route of Hg-mediated health risks [12, 36]. Long-term human exposure to Hg increases its level in blood, sometimes exceeding 150 ng mL^{-1} [37], and this results in negative health risks related to neurodevelopment and neurotoxicity in children and adults [8, 23, 38], immunological, cardiac, motor reproductive and genetic disorders [13], nephrotic syndrome, peripheral neuropathy complications, Alzheimer's, Parkinson's, autism, lupus and amyotrophic lateral sclerosis [39]. Other symptoms may be included such as poor muscle coordination, tingling, numbness in fingers and toes [40, 41], reduced oxidative defense, thrombosis, vascular smooth muscle dysfunction, endothelial dysfunction, and dyslipidemia [33].

7. How to reduce human exposure from mercury sources

- Using clean renewable energy sources rather than the coal;
- Eliminate the use of mercury in mining, gold extraction and other industrial processes;
- Phase out use of non-essential mercury-containing products and implement safe handling, and use and disposal of remaining mercury-containing products [42].

Selenium and fish containing omega-3 fatty acids are thought to diminish mercury toxicity [33] *via* "restoring seleno protein activity" protection against mitochondrial injury and DNA damage, demethylation of methyl mercury and sequestration in complexes, as well as redistribution in the blood away from brain [43].

8. Remediation of Hg-contaminated soils and water

Increasing Hg emissions due to anthropogenic activities caused severe soil pollution issues [44–46]. As a crucial link between the atmosphere and water, soil plays a central role in the global Hg cycle [47]. Soil is not only an Hg sink, receiving Hg input from the environment but also reemitting it to the atmosphere [48, 49], water [42, 50], or the plants grown thereon [39].

Mercury, like other potentially toxic elements, is not biodegradable, and therefore, its remediation should encompass either removal from soil or immobilization [32]. The main Hg removal technologies are physical and chemical remediation methods, as well as bioremediation technology. Adsorption of Hg^{2+} and Hg(0) from water on surfaces of high surface area and high porosity such as chitosan derivatives, synthesized thioether-functionalized covalent triazine nanospheres, pentasil zeolite (type ZSM-5), and utilized silica-coated magnetic nanoparticles are the most common physical approaches for remediating Hg-contaminated soil [51]. Other techniques could help such as soil replacement, physical separation, soil vapor extraction, fixed/stabilized soil, vitrification, thermal desorption, and electrokinetic remediation technology [6]. The latter technique (electrokinetic remediation) depends on passing a direct current between electrodes through the soil to make the Hg ions move through an ion exchange membrane from the soil to the electrodes. The addition of chelating agents to soil could effectively increase the solubility and removal efficiency of Hg [6]. Recently, He and his research group introduced a novel *in situ* immobilization technology by injecting stabilized iron sulfide nanoparticles into soil to immobilize Hg [32].

The *in situ* thermal desorption is a promising technique of Hg remediation that does not need to dig up the contaminated soil; instead, thermal conductive heating (TCH) elements are inserted into the soil in order to directly transfer heat to above 600°C to volatilize various species of mercury, such as HgO, HgS, $HgCl_2$, and mercury associated with organic matter and thus achieve an acceptable decontamination level [51].

Biological remediation/bioremediation depends on plant and microbial in remediation soils [6]. In particular, genetically engineered plants can change methylmercury complexes, and mercury ions into metallic forms of lower toxicity, and then extract, detoxify, and/or sequester this contaminant from soil and water [10]. Phytoremediation is an umbrella term, which refers to the different low cost and eco-friendly technologies that utilize plants in decontaminating areas [52]. This includes: phytostabilization, phytoextraction, and phytovolatilization. This *in situ* application of phytoremediation lessens the disturbance of the surrounding environment and also declines the spread of contamination *via* air and water [32]. There are many other technologies such as the use of nanoparticles to remove/absorb Hg from soil, water, and flue gas, owing to their high adsorption capacity, small dimension, and other unique electrical, mechanical, and chemical properties [53].

Continuous monitoring of Hg levels in air, soils, water, and foods is necessity to ensure their sustainable safe use in order to protect human health and the surrounding ecosystem. In addition, increasing the awareness of humans about the danger of Hg is a proactive step to prevent and reduce the danger of Hg pollution. Furthermore, remediation protocols should be followed in Hg-contaminated areas to lessen its toxicity.

Acknowledgements

This study was supported by the Academy of Scientific Research and Technology, Energy, Environment and Water Nexus (ASRT-APPLE), project number (9112).

Author details

Ahmed A. Abdelhafez[1*], Abdel Aziz Tantawy[2], Mohamed H.H. Abbas[3*], Shawky M. Metwally[4], Amera Sh. Metwally[5], Aya Sh. Metwally[6], Rasha R.M. Mansour[7], Sedky H. Hassan[2,8], Hassan H. Abbas[3], Ihab M. Farid[3], Nermeen N. Nasralla[1], Ahmed S.H. Soliman[1], Mohammed E. Younis[1], Ghada S.A. Sayed[1], Mahfouz Z. Ahmed[1], Ehdaa Alaa Mohamed Abed[1], Ahmed Farouk Al-Hossainy[2], Heidi Ahmed Ali Abouzeid[9], Mahdy H. Hamed[1], Mahmoud I. El-Kelawy[1], Gamal Hassan Kamel[1], Hussein Ferweez[1] and Ahmed M. Diab[1]

1 Faculty of Agriculture, New Valley University, New Valley, Egypt

2 Faculty of Science, New Valley University, New Valley, Egypt

3 Faculty of Agriculture, Benha University, Benha, Egypt

4 Faculty of Agriculture, Zagazig University, Zagazig, Egypt

5 Zagazig University Hospitals, Zagazig University, Zagazig, Egypt

6 Faculty of Veterinary Medicine, Aswan University, Aswan, Egypt

7 Faculty of Specific Education, Benha University, Benha, Egypt

8 College of Science, Sultan Qaboos University, Muscat, Oman

9 Faculty of Pharmacy, New Valley University, New Valley, Egypt

*Address all correspondence to: ahmed.aziz@agr.nvu.edu.eg and mohamed.abbas@fagr.bu.edu.eg

References

[1] Aoyama M. In: Block P, Kasnitz D, Nishida A, Pollard N, editors. Occupying Disability: Critical Approaches to Community, Justice, and Decolonizing Disability. Netherlands: Springer; 2016. pp. 31-45

[2] Ekino S, Susa M, Ninomiya T, Imamura K, Kitamura T. Minamata disease revisited: An update on the acute and chronic manifestations of methyl mercury poisoning. Journal of the Neurological Sciences. 2007;**262**:131-144. DOI: 10.1016/j.jns.2007.06.036

[3] Eto K, Marumoto M, Takeya M. The pathology of methylmercury poisoning (Minamata disease). Neuropathology. 2010;**30**:471-479. DOI: 10.1111/j.1440-1789.2010.01119.x

[4] Yokoyama H. In: Yokoyama H, editor. Mercury Pollution in Minamata. Singapore: Springer; 2018. pp. 53-67

[5] Raj D, Maiti SK. Sources, toxicity, and remediation of mercury: An essence review. Environmental Monitoring and Assessment. 2019;**191**:566. DOI: 10.1007/s10661-019-7743-2

[6] Teng D et al. Describing the toxicity and sources and the remediation technologies for mercury-contaminated soil. RSC Advances. 2020;**10**:23221-23232. DOI: 10.1039/d0ra01507e

[7] Mekawi EM et al. Potential hazards and health assessment associated with different water uses in the main industrial cities of Egypt. Journal of Saudi Chemical Society. 2023;**27**:101587. DOI: 10.1016/j.jscs.2022.101587

[8] Sharma BM, Sáňka O, Kalina J, Scheringer M. An overview of worldwide and regional time trends in total mercury levels in human blood and breast milk from 1966 to 2015 and their associations with health effects. Environment International. 2019;**125**:300-319. DOI: 10.1016/j.envint.2018.12.016

[9] Nagpal N, Bettiol SS, Isham A, Hoang H, Crocombe LA. A review of mercury exposure and health of dental personnel. Safety and Health at Work. 2017;**8**:1-10

[10] Patra M, Sharma A. Mercury toxicity in plants. The Botanical Review. 2000;**66**:379-422. DOI: 10.1007/bf02868923

[11] Magos L, Clarkson TW. Overview of the clinical toxicity of mercury. Annals of Clinical Biochemistry. 2006;**43**:257-268. DOI: 10.1258/000456306777695654

[12] Natasha et al. A critical review of mercury speciation, bioavailability, toxicity and detoxification in soil-plant environment: Ecotoxicology and health risk assessment. Science of the Total Environment. 2020;**711**:134749. DOI: 10.1016/j.scitotenv.2019.134749

[13] Zahir F, Rizwi SJ, Haq SK, Khan RH. Low dose mercury toxicity and human health. Environmental Toxicology and Pharmacology. 2005;**20**:351-360. DOI: 10.1016/j.etap.2005.03.007

[14] Dai Z et al. Assessing anthropogenic sources of mercury in soil in Wanshan Hg mining area, Guizhou, China. Environmental Science and Pollution Research. 2013;**20**:7560-7569. DOI: 10.1007/s11356-013-1616-y

[15] Pacyna JM. Recent advances in mercury research. Science of The Total Environment. 2020;**738**:139955. DOI: 10.1016/j.scitotenv.2020.139955

[16] Rodríguez Martín JA, Nanos N. Soil as an archive of coal-fired power plant mercury deposition. Journal of Hazardous Materials. 2016;**308**:131-138. DOI: 10.1016/j.jhazmat.2016.01.026

[17] Programme, U. N. E. Global mercury assessment. 2018. Available from: https://www.unep.org/explore-topics/chemicals-waste/what-we-do/mercury/global-mercury-assessment

[18] Burns M. Death is not the end. Renew: Technology for a Sustainable Futures. 2022:53-56

[19] Joy A, Qureshi A. Mercury in dental amalgam, online retail, and the Minamata convention on mercury. Environmental Science & Technology. 2020;**54**:14139-14142. DOI: 10.1021/acs.est.0c01248

[20] Tibau AV, Grube BD. Mercury contamination from dental amalgam. Journal of Health & Pollution. 2019;**9**:190612. DOI: 10.5696/2156-9614-9.22.190612

[21] Ball LK, Ball R, Pratt RD. An assessment of thimerosal use in childhood vaccines. Pediatrics. 2001;**107**:1147-1154. DOI: 10.1542/peds.107.5.1147

[22] Mohamed ME-B, El-Meligy MMS, Bushra RR, Mohamed EK. Effect of mercuric chloride exposure during pregnancy and lactation on the postnatal development of the liver in the albino rat. The Egyptian Journal of Anatomy. 2019;**42**:10-27. DOI: 10.21608/ejana.2019.251310

[23] Yawei S, Jianhai L, Junxiu Z, Xiaobo P, Zewu Q. Epidemiology, clinical presentation, treatment, and follow-up of chronic mercury poisoning in China: A retrospective analysis. BMC Pharmacology and Toxicology. 2021;**22**:25. DOI: 10.1186/s40360-021-00493-y

[24] Doshi M, Annigeri RA, Kowdle PC, Subba Rao B, Varman M. Membranous nephropathy due to chronic mercury poisoning from traditional Indian medicines: report of five cases. Clinical Kidney Journal. 2018;**12**:239-244. DOI: 10.1093/ckj/sfy031

[25] Usman S et al. Microplastics pollution as an invisible potential threat to food safety and security, policy challenges and the way forward. International Journal of Environmental Research and Public Health. 2020;**17**:9591

[26] Ravichandran M. Interactions between mercury and dissolved organic matter—A review. Chemosphere. 2004;**55**:319-331. DOI: 10.1016/j.chemosphere.2003.11.011

[27] Lu X, Jaffe R. Interaction between Hg(II) and natural dissolved organic matter: A fluorescence spectroscopy based study. Water Research. 2001;**35**:1793-1803. DOI: 10.1016/S0043-1354(00)00423-1

[28] Liu M et al. Impacts of farmed fish consumption and food trade on methylmercury exposure in China. Environment International. 2018;**120**:333-344. DOI: 10.1016/j.envint.2018.08.017

[29] Šípková A, Száková J, Hanč A, Tlustoš P. Mobility of mercury in soil as affected by soil physicochemical properties. Journal of Soils and Sediments. 2016;**16**:2234-2241. DOI: 10.1007/s11368-016-1420-7

[30] Naharro R, Esbrí JM, Amorós JÁ, García-Navarro FJ, Higueras P. Assessment of mercury uptake routes at the soil-plant-atmosphere interface. Geochemistry: Exploration,

Environment, Analysis. 2018;**19**:146-154. DOI: 10.1144/geochem2018-019

[31] Wang S et al. Accumulation, transfer, and potential sources of mercury in the soil-wheat system under field conditions over the loess plateau, Northwest China. Science of the Total Environment. 2016;**568**:245-252. DOI: 10.1016/j.scitotenv.2016.06.034

[32] He F et al. In situ remediation technologies for mercury-contaminated soil. Environmental Science and Pollution Research. 2015;**22**:8124-8147

[33] Houston MC. Role of mercury toxicity in hypertension, cardiovascular disease, and stroke. The Journal of Clinical Hypertension. 2011;**13**:621-627. DOI: 10.1111/j.1751-7176.2011.00489.x

[34] Abdelhafez AA, Li J. Environmental monitoring of heavy metal status and human health risk assessment in the agricultural soils of the Jinxi River area, China. Human and Ecological Risk Assessment Human and Ecological Risk Assessment. 2015;**21**:952-971

[35] Abdelhafez AA, Li J, Abbas MHH. Feasibility of biochar manufactured from organic wastes on the stabilization of heavy metals in a metal smelter contaminated soil. Chemosphere. 2015;**117**:66-71. DOI: 10.1016/j.chemosphere.2014.05.086

[36] Branco V et al. Biomarkers of mercury toxicity: Past, present, and future trends. Journal of Toxicology and Environmental Health, Part B. 2017;**20**:119-154. DOI: 10.1080/10937404.2017.1289834

[37] Kamensky OL, Horton D, Kingsley DP, Bridges CC. A case of accidental mercury intoxication. The Journal of Emergency Medicine. 2019;**56**:275-278. DOI: 10.1016/j.jemermed.2018.12.039

[38] Carocci A, Rovito N, Sinicropi MS, Genchi G. In: Whitacre DM, editor. Reviews of Environmental Contamination and Toxicology. New York, Dordrecht, London: Springer International Publishing; 2014. pp. 1-18

[39] Khan F, Momtaz S, Abdollahi M. The relationship between mercury exposure and epigenetic alterations regarding human health, risk assessment and diagnostic strategies. Journal of Trace Elements in Medicine and Biology. 2019;**52**:37-47

[40] Matta G, Gjyli L. Mercury, lead and arsenic: Impact on environment and human health. Journal of Chemical and Pharmaceutical Sciences. 2016;**9**:718-725

[41] Caroline MD, Frederic D. Nervous system disorders induced by occupational and environmental toxic exposure. Open Journal of Preventive Medicine. 2012;**2**:272-278

[42] Yan X et al. Recent progress in the removal of mercury ions from water based MOFs materials. Coordination Chemistry Reviews. 2021;**443**:214034. DOI: 10.1016/j.ccr.2021.214034

[43] Spiller HA, Hays HL, Casavant MJ. Rethinking treatment of mercury poisoning: The roles of selenium, acetylcysteine, and thiol chelators in the treatment of mercury poisoning: A narrative review. Toxicology Communications. 2021;**5**:19-59. DOI: 10.1080/24734306.2020.1870077

[44] Daim Q, Feng X, Tang G. The geochemical behavior of mercury in soil and its pollution control. Geologygeochemistry. 2002;**30**:75-78

[45] Hoyer M, Burke J, Keeler G. Atmospheric sources, transport and deposition of mercury in Michigan: Two years of event precipitation. Water, Air,

and Soil Pollution. 1995;**80**:199-208. DOI: 10.1007/BF01189668

[46] Li P, Feng XB, Qiu GL, Shang LH, Li ZG. Mercury pollution in Asia: A review of the contaminated sites. Journal of Hazardous Materials. 2009;**168**:591-601. DOI: 10.1016/j.jhazmat.2009.03.031

[47] Reis AT, Davidson CM, Vale C, Pereira E. Overview and challenges of mercury fractionation and speciation in soils. TrAC Trends in Analytical Chemistry. 2016;**82**:109-117. DOI: 10.1016/j.trac.2016.05.008

[48] Carpi A, Lindberg SE, Prestbo EM, Bloom NS. Methyl mercury contamination and emission to the atmosphere from soil amended with municipal sewage sludge. Journal of Environmental Quality. 1997;**26**:1650-1655. DOI: 10.2134/jeq1997.00472425002600060027x

[49] Liu K et al. Measure-specific effectiveness of air pollution control on China's atmospheric mercury concentration and deposition during 2013-2017. Environmental Science & Technology. 2019;**53**:8938-8946. DOI: 10.1021/acs.est.9b02428

[50] Balogh SJ, Meyer ML, Johnson DK. Transport of mercury in three contrasting river basins. Environmental Science & Technology. 1998;**32**:456-462. DOI: 10.1021/es970506q

[51] Wang L et al. Remediation of mercury contaminated soil, water, and air: A review of emerging materials and innovative technologies. Environment International. 2020;**134**:105281. DOI: 10.1016/j.envint.2019.105281

[52] Jhilta P, Dipta B, Rana A. In: Prasad R, editor. Phytoremediation for Environmental Sustainability. Springer Nature Singapore; 2021. pp. 83-111

[53] Wang Y et al. Green synthesis of nanoparticles for the remediation of contaminated waters and soils: Constituents, synthesizing methods, and influencing factors. Journal of Cleaner Production. 2019;**226**:540-549. DOI: 10.1016/j.jclepro.2019.04.128

Chapter 8

Mercury in the Colombian Caribbean: The Bay of Cartagena, A Model in Resilience

Néstor Hernando Campos and José Luis Marrugo-Negrete

Abstract

The Caribbean Sea in Colombia maybe being subjected to discharges of terrigenous solid waste and with these probably, the natural metallic constituents of the sediments, through the discharges of the Magdalena River since the time of the conquest. With the opening of the Dique canal in the mid-seventeenth century, which linked pipes, swamps, and branches from the Magdalena River to its mouth at the southwestern end of the bay, great changes could be caused from the point of view of mixing the fresh and turbid water of the channel with the clear and marine waters of the bay, which led to the beginning of the deterioration of the coral ecosystems present in the bay. Mercury contamination in the Colombian Caribbean has different origins. Artisanal gold mining has the greatest impact and has generated mercury contamination in many ecosystems, particularly in Bolívar and the Mojana region (department of Sucre and Cordoba). In this study, published information on mercury contamination along the Colombian Caribbean coast is compiled. The concentrations present differences between different areas of the coast. The bay of Cartagena is one of the areas most impacted by this pollutant, mainly due to the discharge of waste from a salt processing plant. Other areas are impacted by mercury, the product of the artisanal exploitation of gold, the discharges as a result of this activity are transported mainly to the Magdalena River, and through the different arms that form the delta, they are deposited in the Colombian Caribbean.

Keywords: mercury, Caribbean, Colombia, marine pollution, biota

1. Introduction

Marine pollution has received great attention a long time ago. In 1959, the first international conference on marine pollution problems was held in Berkley (United States) [1]. This problem is of great importance because the oceans have become the place of deposit and storage of a large part of the waste of all kinds, produced by man. These residues drain into the main river arteries and are transported by this means to their final destination in the marine environment. An important role in the transport of pollutants occurs through the lower layers of the atmosphere, which by precipitation or deposition reach the marine environment.

Mercury is, in very small amounts in seawater, a poison and a danger to life processes. Where an enrichment of this metal occurs, catastrophic consequences occur. It is a ubiquitous environmental toxicant. It exists in three forms, elemental (Hg(0)), inorganic (Hg^{2+}), and organic forms. Hg (0) takes a liquid form at room temperature but readily evaporates into mercury vapor. Hg^{2+} occurs naturally in the environment in the form of divalent cationic salts of mercury, such as HgCl2 and $Hg(OH)_2$, among others. Among the three forms, organic mercury, primarily methylmercury (MeHg), is the most dangerous form.

Methylmercury is a bioavailable form and can bioaccumulate through food webs. Shellfish consumption, especially fish consumption, is the main source of human exposure to MeHg. In 1956, "Minamata methylmercury poisoning" (MPM) was recognized, this being the first incident in the world, although there were some events in which several people had suffered direct health damage from exposure to mercury. organic mercury in a laboratory or factory [2].

According to Yokoyama [2], MPM is a neurological syndrome that was caused by the ingestion of fish and shellfish contaminated by methylmercury compounds, generated in the acetaldehyde production process. The first and second outbreaks of this type of disease in Japan were caused by effluent discharged from a Shin-Nippon Chisso Hiryo (hereinafter referred to as Chisso) factory in Minamata, Kumamoto Prefecture, and a Showa Denko factory in Kanose Town, Prefecture of Niigata, respectively.

MeHg is a stable organic mercury compound and is the most toxic form of mercury in the environment not only for humans but also for wildlife (Wolfe et al. 1998, Henriques et al. 2015, in [2]). Because methylmercury is lipid-soluble, it crosses the blood-brain barrier and accumulates in the brain. Methylmercury in the brain causes lysis of central nervous system cells, resulting in irreversible, permanent cell damage (Rabenstein 1978, in [2]). Therefore, MPM is widely recognized as a disorder in the brain, while Shiraki (1979, in [2]) suggested that MPM produces lesions not only in the brain but also in the vascular and endocrine systems.

In 2005, the European Environmental Bureau and the Mercury Policy Project formed "The Zero Mercury Working Group" (ZMWG), an international coalition of more than 95 public interest non-governmental organizations defending the environment and health. from more than 52 countries. The ZMWG strives to eliminate mercury supply, demand, and emissions from all anthropogenic sources, in order to minimize the presence of mercury in the global environment. Its mission is to advocate and support the adoption and implementation of a legally binding instrument containing the necessary obligations to eliminate, as far as possible, and if not minimize, the global supply and trade of mercury, its global demand, the release anthropogenic release of mercury into the environment, and human and wildlife exposure to mercury, these actions gave rise to the "Minamata Convention" [3].

In the agreement, in article 19, it establishes the research, development, and monitoring of the contaminant, especially with the elaboration of models and the geographically representative monitoring of the levels of mercury and its compounds in vulnerable populations and the environment, including biotic media such as fish, marine mammals, sea turtles, and birds, as well as assessments of the effects of mercury and its compounds on human health and the environment, as well as the social, economic, and cultural effects, especially with regard to vulnerable populations, among others [4].

2. The bay of Cartagena

The bay of Cartagena is located in the middle part of the Colombian Caribbean coastline, between 10°16′–10°26′ N and 75°30′–75°36′ W.

The bay of Cartagena is a semi-enclosed body of water, in which two parts stand out, an external one that connects it with the Caribbean Sea through two mouths (Bocachica and Bocagrande); and the internal one located in the northeastern part and is not directly connected to the sea. The bay of Cartagena has estuary characteristics due to the contributions of continental waters from the Canal del in Pasacaballos.

The first studies focused on the study of mercury contamination in the bay were carried out with the support of the National Institute for the Defense of Renewable Resources—Food and Agriculture Organization of the United Nations (INDERENA-FAO, Acronym in Spanish) [5]. Between 1978 and 1979, a study was carried out with the support of the Oceanographic and Hydrological Research Center of the Colombian Navy (CIOH) to evaluate the dynamics and the chemical and sedimentological characteristics, in the study mercury was analyzed, and eight samplings were carried out quarterly; however, they only present the results of two contrasting periods.

Figure 1 shows the results of the two seasons [6]. According to these authors, in the December–April, dry period, the maximum concentrations were measured in the southwestern part, at stations E37 and E31, in the Mamonal industrial zone and where the Petrochemical and Acalis de Colombia were located, with values between 2.0 and 1.4 μg Hg/l, respectively. In the August–December, rainy seasons, a three-year period, three important foci were detected, one in the Internal Bay (E3) where the

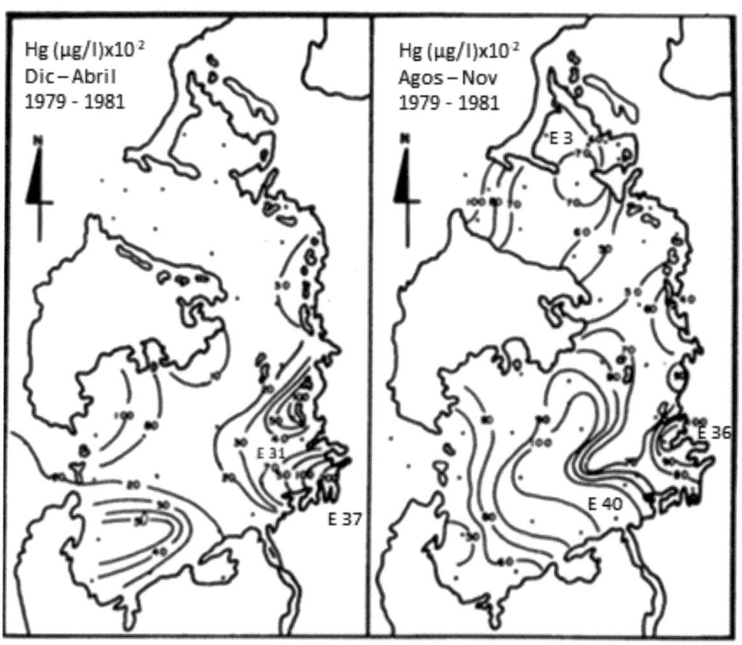

Figure 1.
Mercury content in μg l^{-1} < in water samples at 42 stations in the Bay of Cartagena, in two seasons, December–April and August–November, 1979–1981. Modified from Pagliardini et al. [6].

Los Pegasos dock is located with heavy traffic and possibly strongly influenced during this time of year by the currents, with contents of 1445 μg l^{-1}. The second in Mamonal with contents of 1180 μg l^{-1}, slightly lower than those measured during the windy period. The third is the mouth of the Canal del Dique with a content of 1 μg l^{-1}, which corroborates the importance of the semi-artificial canal in the discharge of sediments and pollutants into the bay. This was corroborated with the horizontal profile that allowed us to observe that the Canal del Dique during the rainy season is the main factor in the distribution of pollutants due to the amount of sediment it transports (see fig 46 and 47, page 86).

In 1996 [7] conducted a study to compare the Hg concentration in two species of estuarine fish, between the Bay of Cartagena and the Ciénaga Grande de Santa Marta, in addition to the content of sediments of these two ecosystems.

The results in sediments (**Figure 2**) clearly show a difference between the stations of the Bahía de Cartagena and the Ciénaga Grande de Santa Marta (CGSM), being the contents measured in the Bay were substantially higher than in the CGSM. The average content for the bay was 1876 ± 578 μg Hg g^{-1} p.s., with extreme values of 94–10,293 μg Hg g^{-1} p.s. A heterogeneous distribution was determined between the different stations of the bay. The maximum values were determined in sediments from station 3, which corresponds to the area of influence of the old chlor-alkali plant, and a decrease in the contents towards the north (stations 3–5) was determined. It is noteworthy that these authors determined the lowest concentrations at station 1 located on the southwestern side, to the right of the mouth of the Clarín Channel, and to the south of the area of influence of the Alcalis discharge, with values of 154 ± 21 μg Hg g^{-1} p.s., which corresponds to station E37, also with the highest values in water published by Pagliardini *et al.* [6].

Compared to the concentrations measured in the Ciénaga Grande (20–109 μg Hg g^{-1} d.w.), they are very low, to the point that the maximum concentration measured is compared to the lowest value in the Bahía de Cartagena (94–10 293 μg Hg g^{-1} d.w.).

Cogua *et al.* [8] determined the contents of total mercury (HgT) and methylmercury (MeHg) in sediments and seston (Suspended organic and inorganic matter) of the Cartagena Bay, in five stations, collected quarterly for one year in 2006. The average content of HgT was 0.18 ± 0.001 μg Hg g^{-1} d.w. The highest contents were measured in front of the industrial zone (0.55 ± 0.03 μg Hg g^{-1} d.w.), with concentrations

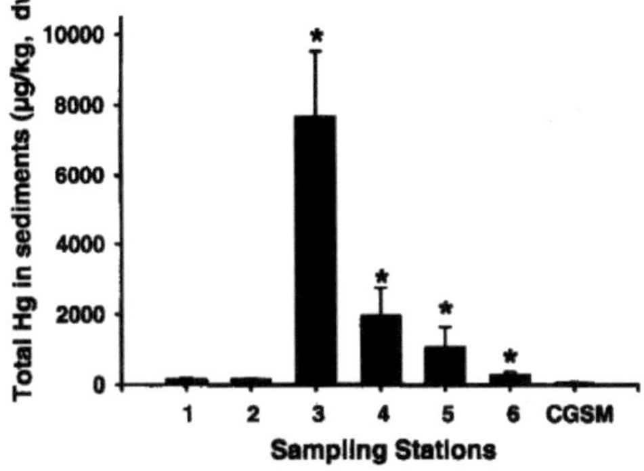

Figure 2.
Average content and standard deviation of total mercury (μg Hg g^{-1} d.w.) in sediments of Bahía Cartagena (six stations) and Ciénaga Grande de Santa Marta (three stations), between March and October 1996. Modified from Alonso et al. [7].

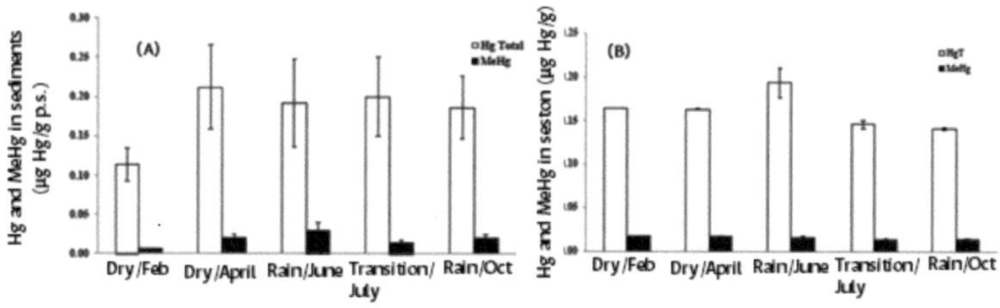

Figure 3.
HgT and MeHg contents (µg Hg g^{-1}) in sediments (A) and seston (B) during the four sampling periods. Modified from Cogua et al. [8].

decreasing towards the north and east. Likewise, it was observed that the highest contents corresponded to the months of influence of the rainy season (April to October), being higher in April at the beginning of the rains. According to these authors, 10% of HgT corresponds to MeHg and they are positively related ($r = 0.87$, $p < 0.04$).

For the contents in the suspended material, a similar behavior was presented, although with a lower average content (0.16 µg Hg g^{-1} d.w.), with the highest concentrations in the rainy season (June 0.19 µg g^{-1} p.s.), and a process of dilution from the station in front of the industrial zone (**Figure 3**).

Although the contents measured in sediments in the study by Cogua *et al.,* [8] are lower than those of Alonso *et al.* [7], the values were always higher in the area of influence of the industrial zone, where the chlorine—alkali was located, which discharged its waste through the Casemiro pipe to the Bay.

3. Organisms

Table 1 shows the mercury content measured in the two species of fish, the mullet and the silver mojarra (*Mugil incilis* and *Euguerres plumieri*), during the sampling period.

The contents in the two species of fish showed large fluctuations, which is to be expected, since they are two resident species in the bay, but they move throughout the study area and are, therefore, subjected to different concentrations for short periods of time. The highest concentrations for the two species occurred during the March sampling followed by November. During the four samplings, the contents were higher in the mojarra. According to the authors, significant differences were determined in the contents ($p < 0.001$), and the differences were 7.3 times greater in *E. plumieri* than

	Mugil incilis		***Euguerres plumieri***	
	X ± SX	**Extremps**	**X ± SX**	**Extremps**
March	87 ± 22	30–166	334 ± 117	37–852
May	10 ± 2	LD–16	160 ± 74	19–582
August	19 ± 10	LD–77	104 ± 22	29–194
November	41 ± 11	LD–89	255 ± 104	43–837

Values are taken from Alonso et al. [7]. X = average value; SX = standard deviation.
The mean value and standard error are given. LOD Limit of detection (7.4 µg Hg g^{-1} < d.w.).

Table 1.
Total mercury content (µg Hg g^{-1} d.w.) in two species of fish from the Bay of Cartagena, collected between March and November 1996.

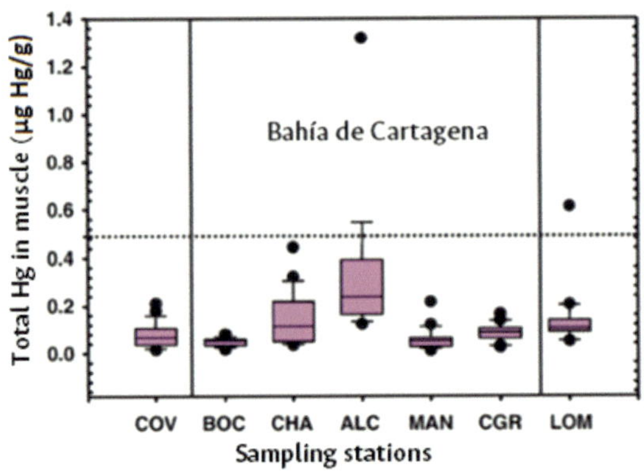

Figure 4.
Total mercury concentrations in muscle of two species of crab of the genus Callinectes in two areas of the Colombian Caribbean, Gulf of Morrosquillo and Bahía Cartagena. COV: Coveñas, BOC: Bocachica; CHA: Puerto Charcoal; ALC: Planta de cloro-alkalis; MAN: Manga; CGR: Castillo Grande; LOM: Lomarena in the Departamento del Atlántico. From Oliveros et al. [9].

in *M. incilis*. These results are understandable, taking into account that the mojarra is a secondary consumer, while the mullet is a detritivore. The contents in the fish from the bay were 7.3 times higher than those from the swamp.

The content of HgT has also been measured in species of crab for human consumption, crabs or swimming crabs (*Callinectes sapidus* and *C. bocourti*) from different collection points (Cartagena and Coveñas), along the Caribbean coast in Colombia [9]. Unfortunately, these authors did not include the HgT content in the crab samples, they only give the values of the percentage of crabs in which Hg was determined.

Figure 4 shows the HgT contents in crab muscle. The highest contents were measured in the specimens collected in the area of influence of the old chlor-alkali plant, followed by those collected in the Charcoal port, on the western side of the mouth of the Clarín Channel. The contents of the crabs collected in the other stations were lower. These results show that despite the chlor-alkali plant having been closed for so long, the presence of Hg in the environment is still notorious.

4. The coastal region of the department of Magdalena

The coastal marine zone of the Department of Magdalena is located between 11°15′33″ N and 73°34′48″ W on the border with the Department of La Guajira and 11°05′42″ N and 74°50′55″W, in Bocas de Ceniza, on the eastern bank of the Magdalena River. There is a great diversity of ecosystems, from sandy beaches, coral reefs, seagrasses, mangroves, rocky coastlines, and sedimentary bottoms, which are considered strategic for the region and the country and provide environmental goods and services that influence the culture. and the economy of the human population of the department, due to its importance in tourism, fishing, and port activities. This region is influenced by the presence of the Sierra Nevada de Santa Marta (SNSM) in which numerous rivers are born and flow into the coastal area of the department and determine the structure of the coast in a series of cliffs and bays.

The outer delta of the Magdalena River is located on the western side of the strip, giving shape to the Ciénaga Grande de Santa Marta (CGSM), the largest lagoon-coastal system in the country and in which the Magdalena River and several rivers from the SNSM interact, and the surrounding seawater.

Even though numerous studies on marine pollution have been carried out on this coastal strip, there are few in which Hg content has been measured. Since 2000, the Marine and Coastal Research Institute [10] has been leading the "Diagnosis and Evaluation of the Quality of Marine and Coastal Waters in the Colombian Caribbean and Pacific" program, with the participation of several government entities (CARs: Autonomous and Regional Corporations), only in the period between June and November 2017 and between February and July 2018, determinations of Hg content in sediments were made [11]. The content of Hg in sediments from 18 points along the coastal strip of the department and during the two sampling periods (rainy 2017 and dry 2108) was determined. The contents fluctuated between values lower than the detection limit of the method (<3.0 ngHg g^{-1} p.s.) up to 496 ngHg g^{-1} p.s. During the rainy season, the contents were low and in nine of the stations, the presence of Hg was not detected, while in the dry season it could be measured in most of the stations. The highest contents were determined in the Bay of Santa Marta and the highest value corresponding to the station in the north of the bay near the cabotage dock of the port of Santa Marta (205 in the rainy season and 296 ngHg g^{-1} d.w., in the dry season) and 192 ngHg g^{-1} p.s. at the south bay station.

5. La Ciénaga Grande de Santa Marta

This physiography comprises the eastern sector of the great delta of the Magdalena River and extends between the same river and the Ciénaga Grande de Santa Marta and up to the foothills of the Sierra Nevada de Santa Marta. On the other hand, this region extends from the Caribbean Sea on the Island of Salamanca, to the Caño Ciego, which later flows into the Fundación River, taking before flowing into it, the name of Caño Schiller.

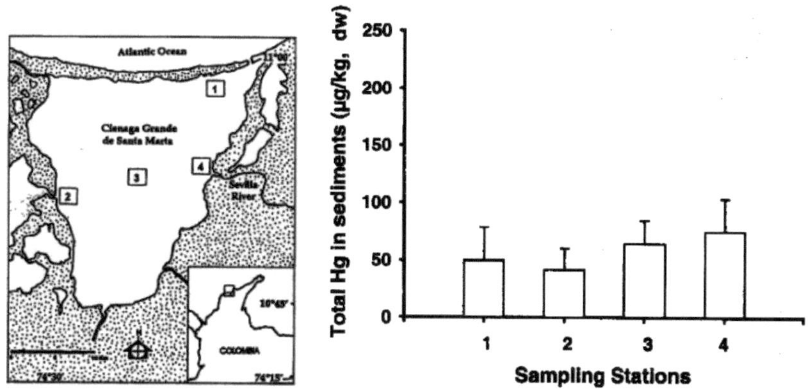

(A) (B)

Figure 5.
(A) Location of the sampling stations in the Ciénaga Grande de Santa Marta. (B) Average content and standard deviation of total mercury (µg Hg g^{-1} d.w.) in sediments of the Ciénaga Grande de Santa Marta (four seasons), between March and October 1996. 1. Islas del Rosario. 2. Boca de Caño Grande, 3. Centro, 4. Boca del Río Sevilla. Modified from Alonso et al. [7].

In the REDCAM program [10], six stations were included in the area of influence of the CGSM, apparently, they were only measured in the dry season and the contents fluctuated between 8.0 and 89.4 ng Hg g^{-1} d.w. The maximum value was measured in front of the mouth of the Sevilla River in the central part of the eastern side (**Figure 5A** Station 4), which drains from the SNSM, followed by the sediments collected near the mouth of the Fundación River, which, in addition to being born in the SNSM, receives the discharges of the Magdalena River, followed by the concentrations determined in sediments from other rivers of the SNSM fluvial system.

According to Alonso *et al.* [7], the Hg content in sediments fluctuated between 20 and 109 μg Hg g^{-1} d.w., with an average value of 58±6. These values are close to natural values (**Figure 5**). Although the authors did not determine significant differences, a slightly greater influence on Hg discharge is observed at the mouth of the Sevilla River, which belongs to the SNSM water system, which is consistent with the REDCAM results.

6. Organisms

Alonso *et al.* [7] determined Hg contents in sediments and fish (*Eugerres plumieri* mojarra and *Mugil curema* mullet) in the CGSM.

E. plumieri is a resident species in the CGSM, a secondary consumer, predominantly of the bivalve *Mytilopsis sallei* that forms banks on the bottom of the swamp, a filter-feeding species, while the mullet (*M. incilis*) is purely detritivorous. In the first, the contents were always about twice as high as in the second. The highest contents corresponded to the samples at the beginning of the year, corresponding to the end of the dry season and the beginning of the rainy season. As mentioned for the bay of Cartagena, the contents in the two species were 7.3 times lower than in BC (**Table 2**).

Campos [12] made the first determinations of Hg in the oyster Crassostrea rhizophorae, in samples collected in January 1987 at two stations located in the northern fringe, one on the western side at the mouth of the Clarín channel that transports and discharges water from of the Magdalena River and the second in the eastern part of the strip, near the mouth of the CGSM in the sea. The contents fluctuated between 0.04 and 0.18 μg Hg g^{-1} p.s., the latter corresponding to the samples from the Clarín channel.

	Mugil incilis		***Euguerres plumieri***	
	X ± SX	**Extremps**	**X ± SX**	**Extremps**
March	15.7 ± 7	LD–51	26 ± 8	LD–68
May	10 ± 2	LD–17	28 ± 6	13–52
August	8 ± 2	LD–17	9 ± 3	LD–22
November	6 ± 1	LD–10	12 ± 3	LD–20

Values taken from Alonso et al. [7]. X = average value; SX = standard deviation.
The mean value and standard error are given. LOD Limit of detection (7.4 μg Hg g^{-1} d.w.).

Table 2.
Total mercury content (μg Hg g^{-1} d.w.) in two species of fish from the CGSM collected between March and November 1996.

Species of fish	Sample	Hg
Eugerres plumieri	Muscle	18.7 ± 5 μg/kg
Cathorops mapale		18.1 ± 14.6 μg/kg
Centropomus undecimalis		28.2 ± 10.1 μg/kg
Elops smithi		36.6 ± 44.0 μg/kg
Eugerres plumieri		13.1 ± 11.3 μg/kg
Mugil incilis		16.1 ± 21.4 μg/kg

Values taken from Pinzón-Bedoya et al. [13].

Table 3.
Total mercury content (μg kg^{-1} fresh weight) in fish species from the CGSM collected between January and December 2018.

Pinzón-Bedoya *et al.* [13] determined the concentrations of Hg (μg/kg f.w, fresh weight) contained in the muscle of fish species usually consumed by inhabitants of the area (**Table 3**). The compilation of the data shows that the results presented by Alonso *et al.* [7] an increase in the concentrations of Hg has been presented, which could demonstrate that the contributions of this metal persist, whether due to anthropic or natural origin, to the ecosystem; Likewise, it is possible to affirm the existence of accumulative and magnifying processes of Hg from the sediments to fish and bivalves.

Alonso *et al.* [7] indicated that the concentrations of Hg measured in the ecosystem components of the Ciénaga Grande de Santa Marta could be attributed to atmospheric deposition processes since this area did not present a significant industrial development. However, Caballero-Gallardo *et al.* [14] warn that the CGSM receives contaminants from intensive agriculture, especially bananas and African palm. Regardless of the source, the truth is that in the results of Pinzón-Bedoya *et al.* [13] Hg concentrations were positively correlated with morphometric variables (weight and length), evidencing Hg bioaccumulation processes in the aquatic biota of this ecosystem.

7. Cispatá Bay

For this coastal ecosystem, Burgos-Núñez *et al.* [15] reported that the bay of Cispatá-Colombia has been affected by chemical fertilizers, herbicides, pesticides, domestic wastewater, and spillage of substances such as hydrocarbons and heavy metals; These reasons were enough for them to determine the polycyclic aromatic hydrocarbons and heavy metals present in the tropical marine ecosystem of this bay (**Table 4**).

The results allowed us to observe that Hg concentrations increased with the trophic level in environments with very low levels of this metal, indicating bioaccumulation in the Cispatá Bay ecosystem. The distribution of Hg in the food web was: sediments < fish < birds. Seabirds can tolerate high concentrations of Hg, due to their ability to demethylate Hg in the liver and then store it as inorganic Hg [15, 16]. Lucia *et al.* [17] pointed out that the feathers in birds serve to excrete excess metals such as Hg; However, Tsipoura *et al.* [18] assured that concentrations higher than 5.0 μg Hg g^{-1} could impair the reproductive performance of birds.

Type of sample	Species of Fish	Sample	Hg
Sediments			31.7 ng/g d.w. (1.63–135.6)
Fish	*Cetengraulis edentulus*	Muscle	0.10 μg/g f.w.
	Eugerres plumieri		0.14 μg/g f.w.
	Centropomus undecimalis		0.38 μg/g f.w.
	Trichirus lepturus		0.67 μg/g f.w.
Birds		Blood	0.23 ± 1.09 μg/L
	Pelecanus occidentalis	Feathers	Juveniles (1.77 ± 0.71 mg/kg f.w) Adults (5.15 ± 1.52 mg/kg f.w)
	Phalacrocorax brasilianus		Juveniles (1.76 ± 0.65 mg/kg f.w) Adults (4.99 ± 1.47 mg/kg f.w)
	Fregata magnificens		Juveniles (2.10 ± 1.36 mg/kg f.w) Adults (10.2 ± 4.99 mg/kg f.w)
	Thalasseus maximus		Juveniles (0.96 ± 0.46 mg/kg f.w) Adults (3.57 ± 1.37 mg/kg f.w)

Values taken from Burgos-Núñez et al. [15].

Table 4.
Concentrations of THg in sediments, fish, and birds of the Bay of Cispatá-Colombia.

8. Mercury and malformation in crabs

In two recent studies, the impact that Hg can have on marine organisms was evaluated. The first is related to the high incidence rate of malformation in crabs collected in Cispatá Bay [19]. According to these authors, malformations occurred throughout the sampling year, fluctuating between 2.78% and 25.09% (**Table 5**).

Figure 6 shows the malformations of two specimens of the Xanthoidea superfamily, multiple malformations are observed in both. In (A) malformations in the two anterolateral borders. (B) The left side shows the separated teeth, the fifth reduced, the third subquadrate projected posteriorly, and the first and second fused. (C) The

	06/05		09/05		12/05		03/06	
	No. D	% D	No. D	% D	No. D	% D	No. D	% D
C. Salado	0	0	9	9.68	0	0	0	0
C. Mocho	2	10.91	0	0	0	0	3	4.11
P. Nisperal	69	25.09	13	6.34	8	2.78	28	11.62
% Total		55.8		15.94		5.8		22.46

Information taken from Campos et al. [19].

Table 5.
Number of crabs of the superfamily Xanthoidea with deformities (No. D) and percentage in relation to the number of individuals collected (% D) of the superfamily, in three stations of Cispatá Bay, in four seasons of one year.

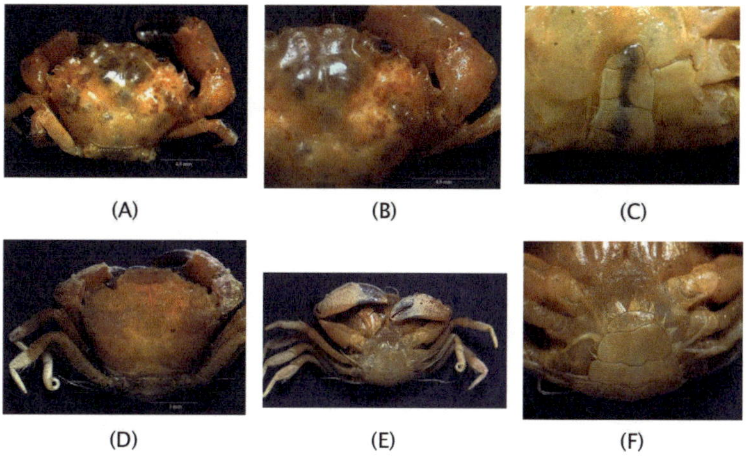

Figure 6.
Two examples of multiple malformations in crabs. (A) Dorsal view of a crab of the Xanthoidea superfamily. (B) Approach to the malformation of the right anterolateral border of the specimen. (C) Malformations in the abdomen. (D) Dorsal view of another crab of the superfamily Xanthoidea, the formation of the dactyl of the third left pereopod is observed. (E) Abnormal growth of the pleopods. (F) Malformation of the abdomen [19].

deformed abdominal segments do not cover the abdominal cavity; the telson ends in a subtriangular shape, the right edge longer than the left. (D) Malformation of the dactyl of the third left pereopod and of the orbits and teeth of the right anterolateral border. (E) The right sternites, fifth and sixth, separated, and those of the left side completely fused. (F) There is an overgrowth of the appendages of the abdomen (pleopods) protruding and the abdomen is twisted to the right due to shortening of the right border of the sixth abdominal segment (**Figure 6**).

The highest incidence of malformations occurred during the sampling of June, with 55.8% of the total malformations and corresponds to the period immediately after the first rainy season (April-May). While the lowest incidence was determined in December, in the full dry season. In Punta Nisperal, crabs with malformations were collected throughout the three sampling periods and with a high prevalence (25.9% in June and 11.62% in March 2006). December was the only season in which specimens with malformations were captured, although with the lowest incidence (2.78%).

9. Methylmercury in sharks

In a recent study Rueda *et al.* [20], evaluated the methylmercury content (μg MeHg kg^{-1} w.w.) in muscle and stomach content of the Antillean dogfish shark (*Rhizoprionodon porosus*), in three different areas in the Colombian Caribbean. The first in Cabo de la Vela north of the Caribbean coast, scarcely intervened by man and influenced by the upwelling of the same name. The second is in the Las Flores station, near the mouth of the Magdalena River, and the third is in the southern part of the Caribbean on Isla Fuerte.

Figure 7 shows the MeHg contents in the dogfish shark. The highest contents were determined at the Las Flores station (B) and the lowest at Cabo de la Vela (A). Regarding the sampling periods, it was observed that the maximum values occurred in the rainy season in Las Flores and in the dry season in Isla Fuerte and Cabo de la

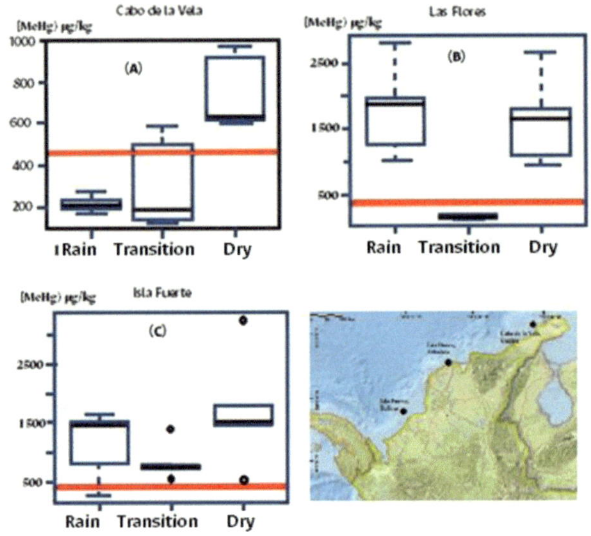

Figure 7.
MeHg content in µg kg^{-1}(w.w.) in Rhizoprionodon porosus muscle according to the season and at each of the study sites. Due to the values obtained, the Y axis of the graph belonging to Cabo de la Vela presents a different scale than the others. Red line indicates the maximum concentration recommended by the WHO (2000) for food consumption. Modified from Rueda et al. [20].

Vela. The lowest values in MeHg concentrations in shark tissues occurred, in general, during the transition period (July to August). For the three samplings in Isla Fuerte and in Las Flores in the dry and rainy seasons and only in the dry season in Cabo de la Vela, the contents exceeded the limits given by the World Health Organization (WHO, 2000) as permissible for the daily consumption per person. It is probable that the intensification of the upwelling by the trade winds, which are stronger, has transported the Hg from the sediments (**Figure** 7).

Figure 8 shows the concentrations of MeHg in µg kg^{-1}(w.w.) in the stomach content of *Rhizoprionodon porosus* according to the season and at each of the study sites. Only one sample was presented in a single sampling site in which the contents exceeded the permissible limits for daily human consumption (500 µg kg^{-1} WHO, 2000). The lowest concentrations were measured at Cabo de Vela, while the maximum was measured at Las Flores station in the dry season (408.6 µg k^{-1} g w.w.), while in the remaining samples, the contents were less than 200 µg/kg w.w., additionally and

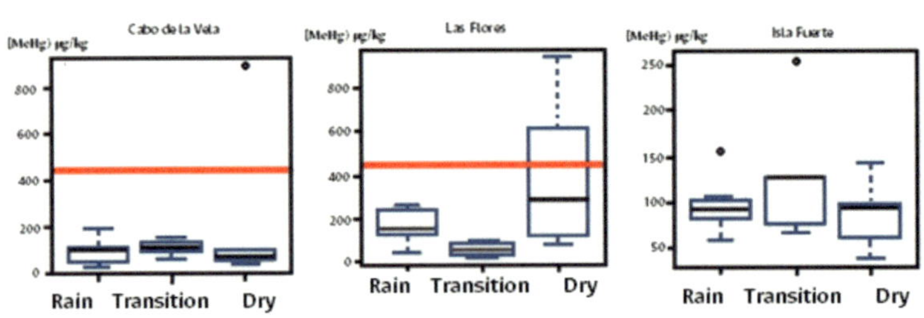

Figure 8.
MeHg concentrations in µg/kg(w.w.) in the stomach content of Rhizoprionodon porosus according to the season and in each of the study sites. The red line indicates the maximum concentration recommended by the WHO (2000) for food consumption. Modified from Rueda et al. [20].

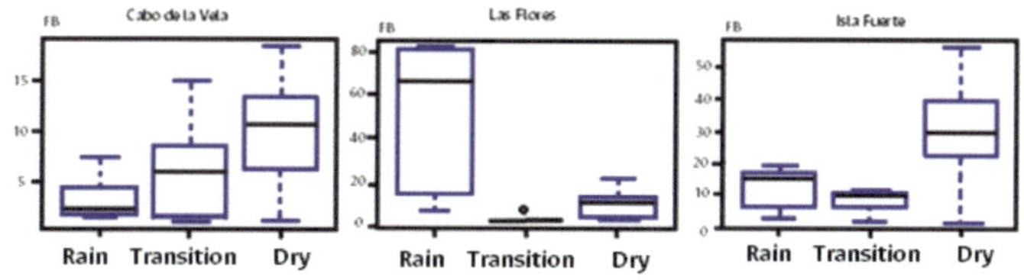

Figure 9.
Biomagnification factor (FB) in Rhizoprionodon porosus according to the season in each of the study sites. Modified from Rueda et al. [20].

with the exception of the flowers in the dry season, there were no significant differences between the samplings and the seasons.

Based on the MeHg concentrations in the muscle and stomach contents of the sharks, the biomagnification factor was determined for each of the sampling sites (**Figure 9**). Regarding the total data from the Colombian Caribbean, the Kruskal-Wallis test showed significant differences ($p = 0.05$) between the values of this factor in the different seasons. However, when doing the analysis with the data from each site separately, the only place where there were significant differences between the seasons was the Las Flores station. In Cabo de la Vela and Isla Fuerte, the temporal variation was not significant, although when analyzing the average values of each site in their respective seasons, they coincided in terms of the notable increase during the dry season.

10. Conclusions

Due to its ubiquity, mercury can be present in a wide range of spaces and environmental matrices that, due to its presence, can alter the balance in aquatic and terrestrial ecosystems. The concentrations, reported in the different studies addressed for this compilation, reveal the diversity of the possible sources for the marine environment of the Colombian Caribbean, generating a cause for concern due to the connotations that this metal has for human and environmental health.

Despite the limitations in accessing financial and logistical resources to carry out each of these investigations, the scientific community has been able to join efforts to show the current status of mercury contamination in coastal ecosystems, understanding that it is necessary to continue with the efforts to understand or elucidate in greater depth the impacts that mercury is having on marine-coastal organisms and on the human population of the Caribbean region. In this sense, the environmental control and surveillance authorities must seek mechanisms to control mercury emission or contamination sources for the protection of marine ecosystems, as established in article 12 of the Minamata Agreement. Also, the same agreement in its article 19, establishes the need to carry out monitoring studies in the environment and in the vulnerable population for the evaluation of the impacts on ecosystems [4].

Contribution No. of the Institute for Studies in Marine Sciences, CECIMAR, of the National University of Colombia, Caribbean Campus.

Author details

Néstor Hernando Campos[1*] and José Luis Marrugo-Negrete[2]

1 Institute of Studies in Marine Sciences – CECIMAR, National University of Colombia, Caribbean Campus, Bogotá, Colombia

2 Universidad de Córdoba, Colombia

*Address all correspondence to: nhcamposc@unal.edu.co

References

[1] Gerlach SA. Meeresverschmutzung. Berlin: Springer Verlag; 1976. p. 145

[2] Yokoyama H. Mercury Pollution in Minamata. Singapore: Springer Briefs in Environmental Science; 2018. p. 74

[3] Lennett D, Gutiérrez R. Convenio de Minamata sobre el Mercurio Manual de ratificación y aplicación. Natural Resources Defense Council y BAN Toxics. 2015

[4] UNEP (United Nations Environment Programme). Minamata Convention on Mercury-Text and Annexes. 2017

[5] Escobar JJ, Granados J, Polensky E, Valle V. Estudio de la contaminación por mercurio en la Bahía de Cartagena. Colombia: Invest; 1977. p. 39

[6] Pagliardini JL, Gómez MA, Gutierrez H, Zapata SI, Jurado A, Garay JA, et al. Síntesis del proyecto Bahía de Cartagena. Boletín Científico CIOH. 1982;**4**:49-110

[7] Alonso D, Pineda P, Olivero J, González H, Campos NH. Mercury levels in muscle of two fish species and sediments from the Cartagena Bay and the Ciénaga Grande de Santa Marta, Colombia. Environmental Pollution. 2000;**109**:175-163

[8] Cogua P, Campos NH, Duque G. Concentración de mercurio total y metilmercurio en sedimento y seston de la Bahía de Cartagena, Caribe colombiano. Bol. Invest. Mar. Cost. 2012;**41**(2):267-285

[9] Oliveros Verbel J, Johnson-Restrepo B, Baldiris-Avila R, Güette-FernándezJ,Magallanes-CarreazoE, Vanegas-Ramírez L, et al. Human and crab exposure to mercury in the Caribbean coastal shoreline of Colombia: Impact from an abandoned chlor-alkali plant. Environmental International. 2008;**34**:476-482

[10] INVEMAR. Diagnóstico y evaluación de la calidad de las aguas marinas y costeras en el Caribe y Pacífico colombianos. Luisa F. Espinosa y Ostin Garcés (Eds). Red de vigilancia para la conservación y protección de las aguas marinas y costeras de Colombia – REDCAM: INVEMAR, MinAmbiente, CORALINA, CORPOGUAJIRA, CORPAMAG, CRA, CARDIQUE, CARSUCRE, CVS, CORPOURABÁ, CODECHOCÓ, CVC, CRC y CORPONARIÑO. Informe técnico 2018. Serie de Publicaciones Periódicas No. 4 del INVEMAR, Santa Marta. 2019

[11] INVEMAR. Diagnóstico y evaluación de la calidad de las aguas marinas y costeras en el Caribe y Pacífico colombianos. In: Espinosa LF, Garcés O, editors. Red de vigilancia para la conservación y protección de las aguas marinas y costeras de Colombia – REDCAM: INVEMAR, MinAmbiente, CORALINA, CORPOGUAJIRA, CORPAMAG, CRA, CARDIQUE, CARSUCRE, CVS, CORPOURABÁ, CODECHOCÓ, CVC, CRC y CORPONARIÑO. Informe técnico 2018. Santa Marta: Serie de Publicaciones Periódicas No. 4 del INVEMAR; 2019. p. 212

[12] Campos C, N. H. La contaminación por metales pesados en la Ciénaga Grande de Santa Marta, Caribe colombiano. Caldasia. 1990;**16**(77):231-244

[13] Pinzón-Bedoya C, Pinzón-Bedoya M, Pinedo-Hernández J, Urango-Cárdenas I, Marrugo-Negrete J. Assessment of

potential health risks associated with the intake of heavy metals in fish harvested from the Largest Estuary in Colombia. International Journal of Environmental Health Research and Public Health. 2020;**17**:1-13

[14] Caballero-Gallardo K, Alcala-Orozco M, Barraza-Quiroz D, De la Rosa J, Olivero-Verbel J. Environmental risks associated with trace elements in sediments from Cartagena Bay, an industrialized site at the Caribbean. Chemosphere. 2020;**242**:125173

[15] Burgos-Núñez S, Navarro-Frómeta A, Marrugo-NegreteJ,Enamorado-MontesG, Urango-Cárdenas I. Polycyclic Aromatic Hydrocarbons and Heavy Metals in the Cispata Bay. Colombia: Marine Pollution Bulletin; 2017

[16] Burger J, Gochfeld M. Effects of Chemicals and Pollution on Seabirds, Biology of Marine Birds. Boca Raton, FL: CRC Press; 2001. pp. 485-526

[17] Lucia M, Andre J, Gontier K, Diot N, Veiga J, Davail S. Trace element concentrations (mercury, cadmium, copper, zinc, lead, aluminium, nickel, arsenic, and selenium) in some aquatic birds of the southwest Atlantic coast of France. Archives of Environmental Contamination and Toxicology. 2010;**58**:844-853

[18] Tsipoura N, Burger J, Newhouse M, Jeitner C, Gochfeld M, Mizrahi D. Lead, mercury, cadmium, chromium, and arsenic levels in eggs, feathers, and tissues of Canada geese of the New Jersey Meadowlands. Environmental Research. 2011;**111**:775-784

[19] Campos-Campos NH, Dueñas-Ramírez PR, Genes N. Malformación en cangrejos de la superfamilia Xanthoidea (Crustacea: Brachyura) en la bahía de Cispatá (Córdoba, Colombia). Revista de la Academia Colombiana de Ciencias Exactas, Físicas y Naturales. 2015;**39**(150):91-99

[20] Rueda-Bernal R, Acero P, et al. Determinación del rol del tiburón cazón antillano *Rhizoprionodon porosus* (Carcharhinidae) en el flujo de metilmercurio en las redes tróficas del Caribe colombiano. Rev. Acad. Colomb. Cienc. Ex. Fis. Nat. 2020;**44**(170):169-181

Chapter 9

Mercury Contamination and Spill-Over at Human-Wildlife-Environment Interface

Andrew Tamale, Justine Okello and Celsus Sente

Abstract

Man's quest for energy demands that fuel for running machines and cooking is vital for mankind. Oil and coal have served this energy quest for time immemorial. This oil quest has been present in the Albertine Graben since 1920, threatening biodiversity spots, terrestrial wildlife, and aquatic resources. The current book chapter provides insights into the spatial distribution of potentially toxic elements (Mercury) in terrestrial and marine species and the health risk posed to terrestrial and aquatic species due to oil exploitation.

Keywords: oil spill, mercury, Lake Albert Graben, human-wildlife environment interface, developing country

1. Introduction

Pollution is a resultant detrimental effect on the ecosystem attributed to hazard discharges and mismanagement [1, 2]. Contamination on the other hand is attributed to presence of hazards in the biological systems and these act as sources of toxicity to flora and fauna [3]. Potential toxic elements (PTEs), another type of inorganic pollutant, make up one of the most significant groups of pollutants that are released into environment by chemical and allied industries such as pharmaceuticals, fertilizers, and refineries. These PTEs can pollute the environment if their concentration is higher than the allowable limits [4]. There is a possibility that these inorganic pollutants cannot be broken down by biological processes, and as a result, they remain in the environment for a while, ultimately causing disruptive effects, not only on public health but also on the flora and fauna of aquatic and terrestrial environment [5, 6]. Potentially toxic elements are a threat on a global scale since they are not biodegradable and have a propensity to build up in living organisms through the food chain. Different body organs in both humans and animals are impacted by a number of potentially toxic substances' causing acute and chronic toxic effects [1, 6]. In many Sub-Saharan African countries, anthropogenic activities that occur within an ecosystem, as well as the natural sources, are the two main contributors to PTEs pollution.

The vast majority of pollution is caused by human activities, such as the search for new sources of energy [7, 8]. Since the beginning of time, coal and oil have been sources of energy for a wide variety of applications across the globe.

In Uganda, oil exploration in the Lake Albert Graben that began as early as 1938 in Butiaba, Buliisa with geological explorations, has today evolved into the construction of an oil pipeline [9]. Such pipelines come along with the threat of pipeline accidents where oil spills are eminent [10]. The Uganda environmental policy stipulates that oil spills are one of the disasters whose level of contamination of the environment would warrant attention [9]. According to Kassim [2], oil resources should be managed in cognition of the negative effects such as pollution and contamination. The environmental policy stipulates that oil spills are one of the disasters whose level of contamination of the environment would warrant attention [11]. The pollution cited in this situation is that of potentially toxic elements, especially mercury and lead. Since "oil spill" is often a new phenomenon in the area, the ecosystems should be able to adapt in a compatible way otherwise the ecosystem collapses. The theory of sustainable utilization for oil resources in the Lake Albert Graben focuses on three main areas namely; availability of the resources, adaptability and flexibility, and homeostasis [12]. The resources targeted at the Albertine Graben include water, assets, and entitlements, and not so much of PTEs research [13]. For instance, much of the work executed about water levels of mercury, levels of mercury in fish, and risk burden thereafter, is documented better in developed countries compared to the developing countries, like Uganda. The use of consumption advisories is where most of the information in developed countries is stored. The assets in the biodiversity hot spot include flora and fauna uniquely found in this site [14]. In addition to the animal species and plants in the protected area, lies a whole new industrial park as part of the oil city. The much-anticipated livelihood from this oil city creates an entitlement to the community around the oil drilling area [14]. However, conflicts might result in negative catastrophes, such as oil leaks, if the delicate balance between entitlement and livelihood is not realized. In the Murchison Falls Conservation Area (MFCA), wildlife, domestic animals, and human ecosystems are documented as resilient systems, initiated by anthropogenic activities [15]. The presence of oil and gas prospecting in MFCA as major anthropogenic activities translates into ecosystem resilience, coping, and tolerance as a result of the possible oil spills [15]. Oil and gas production is one of the precursors of mercury in the environment, should there be an oil spill [16]. The surrounding communities are being empowered through a socioecological approach that focuses on sustainability of the ecosystem proponents.

This chapter, therefore, discusses the detrimental effects of mercury on an ecosystem that lies at the wildlife, domestic animal, and human interface. These effects are on around an oil spill framework that encompasses terrestrial and aquatic environments, species, and habitats as illustrated in **Figure 1**.

The likely exposure factors include exposure time, exposure quantity, and analytic chemicals, that is, mercury and lead. The species-related factors include species identity, development stage, generation time, feeding mode, and mobility. The habitat related factors include depth at which species thrive, environmental stressors, i.e., deforestations, erosion, anthropogenic activities and climatic conditions. These triad factors also are both inter and intra linked.

Location and habitat.

In the Lake Albert region, oil exploration often excites the population. However, the location of oil wells in MFCA biodiversity hot spots and forests has led to loss of habitat and ecosystem for biodiversity. The location of MCFA has been illustrated in **Figures 2** and **3**.

Figure 1.
Framework of oil spill burden on Ecosystem. Sources: [17, 18].

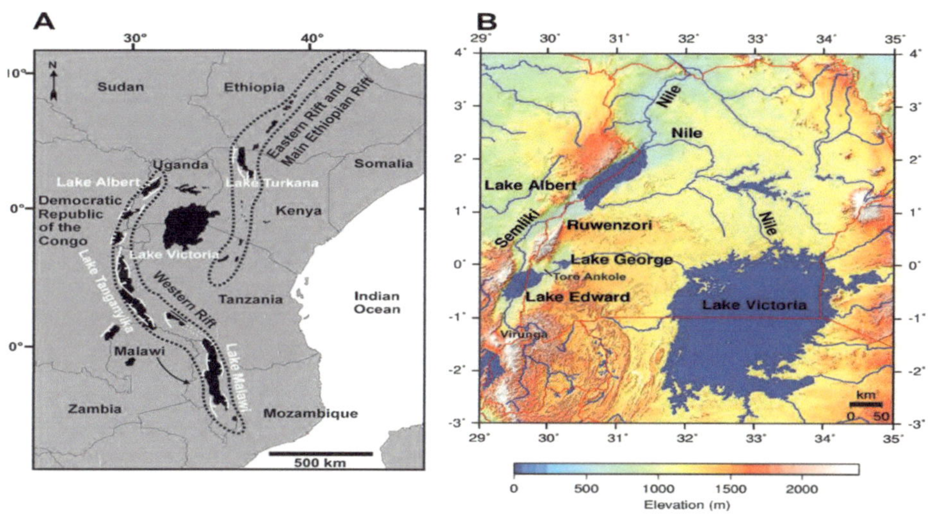

Figure 2.
Map of Lake Albert region [19].

Regardless of where the oil spill occurs, whether terrestrial or aquatic, the impacts are devastating, as they can lead to accumulation of potentially toxic elements, especially mercury in the terrestrial and aquatic environment, causing toxicities in animals and humans. Mercury in particular bioaccumulates and is magnified along trophic levels, often bioconcentrate in plants and animals and the flora acts as sources of the potentially toxic substances/mercury.

On the other hand, there have been positive impacts on development, that is, infrastructure, road networks, and improved livelihoods of the surrounding communities. A balance is needed in utilization of the oil resource since it is perishable but can also have vast negative impacts, which include the oil spills.

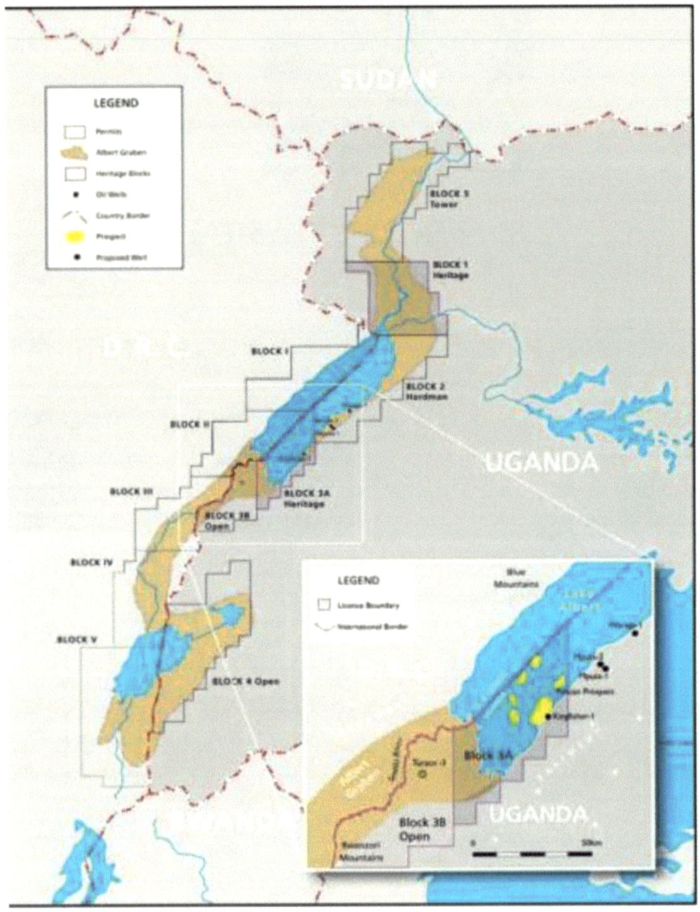

Figure 3.
Oil fields in Uganda [20].

2. Species

The conservation world is aware of the burden of oil spills in the areas where oil exploration and extraction exist. These oil spills mainly negatively affect the wildlife (flora and fauna), fishing communities, and the communities in and around the wildlife-protected areas. To safeguard these vulnerable populations at the animal-environment-human interface, solutions on how to mitigate the levels of potential toxic substances found in water, plants, and animals need to be found. This information is displayed as a distribution map of the key levels of lead and mercury in food eaten by communities in the protected areas, and associated health risks. The adverse effects of lead and mercury on wildlife, aquatic species, and environment also need to be documented. These documents translate into the food consumption advisories for lead and mercury in the MFCA. The awareness of risks and benefits associated with eating fish, plants, and wildlife from MFCA is vital for tourists and communities around the protected area. Since most of the mercury is taken in through fish consumption, vulnerable populations should watch out for the fish species eaten, especially the carnivorous fish. The vulnerable populations include children less than 12 years, women of childbearing age, and elderly.

DOI: http://dx.doi.org/10.5772/intechopen.109492

3. Anthropogenic activities

Anthropogenic activities offer a delicate balance between the needs of a community/government and the biodiversity spots and their contribution to nature [21]. The country's development process should be in tandem with the environmental concerns [20]. Therefore, activities like oil exploration in biodiversity spots should be allowed to execute with guiding regulations and policy. If the works go on unregulated, the destruction to the environment and the associated costs of restoration are quite prohibitive. The cost of an oil spill to wildlife range from losses due to acute toxicity of the aquatic and terrestrial species, the cost of rehabilitation of those that can be recovered, and the cost of clean up to and the length of 1300 km length of the oil pipe for crude oil. The work on oil spills is much needed in the Lake Albert Graben given the fact that 4 million gallons per day in the kingfisher project shall be produced come 2023. The Lake Albert Graben has over 7 blocks of which the kingfisher project lies in one of the blocks [2, 22]. The Lake Albert Graben is synonymous with the Niger delta in Nigeria. Oil-related studies in Nigeria exhibited a range of between 230 and 1200 spills per year in the Niger delta between 1981 and 2015 [3]. The amounts per oil spill in the oil mining areas and between the path of oil pipelines can only be estimated to be between 50,000 barrels and 300,000 barrels per year in the same study period in Niger delta [3]. Based on the oil spills, fish are contaminated with potentially toxic substances and this not only leads to health risks in high-level trophic fish but also has a negative effect on fish breeding in the water sources. Adding the restoration costs to the 3.5 billion US$ initial cost of construction of the pipeline is a tall order [3].

3.1 Hazards

Pollution attributed to quest for energy in Uganda and world over is related to cancers and other severe health burdens to animals and humans. Some of the detrimental hazards associated with oil spills are potentially toxic substances, especially mercury. Considering the Lake Albert Graben where oil wells are located both inland and in lake water, aquatic fish and captive animals will be exposed to levels of Mercury exceeding the maximum allowable limits of WHO and FAO. Exposure of animals and aquatic species to Mercury results in neurotoxicity, kidney damage, cancers, and teratogenicity [6, 10, 23]. One of the gaps identified in the study is the absence of an investigation into the establishment of potentially toxic substances levels in plants, wildlife, and aquatic species, establishment of the associated health risks, and lack of the PTEs distribution map for the MFCA.

The hazard identification study employs teams of epidemiologists, researchers, wildlife veterinarians, game park staff, statisticians, lab specialists, and policy analysts. The study approach utilizes quantitative approach of research involving semi-structured interviews, sample collection, lab investigation, data mining, and dissemination of findings/report/policy and these translate into risk assessments. The evidence generated is what is utilized to reduce oil pollution effects on wildlife and aquatic species in MFCA. Efforts like these steer the reduction of pollutants in areas where biodiversity is paramount and sustainable management of the ecosystem is warranted. Areas that have oil wells and parks in Uganda and other developing countries can utilize study findings to design appropriate interventions for reduction in the negative effects of oil spills in the MFCA. The resultant interventions will result in ecosystem recovery.

3.2 Current literature

Pollution attributed to expedition for energy in Uganda and world over is closely linked to several health burdens to animals, humans, and environment [24]. This evidence can be much more pronounced in the potentially toxic substances, which have been documented to cause cancers in animals and human populations [19, 23, 25]. Furthermore, the plants have equally been shown to contain traces of potentially toxic elements in amounts beyond acceptable levels [23]. It is imperative to note that some of the detrimental hazards associated with extraction and refining of petroleum and associated products are the oil and potentially toxic elements spill-overs [3, 26]. Mercury and its compounds are one such PTEs that is linked to deleterious effects in the nervous system, kidneys, and liver and disturb immune response processes, causing tremors, impaired vision and hearing, paralysis, and emotional instability [27, 28]. Mercury is one of the most common contaminants associated with oil spills [29]. Methods of biological control where eco-friendly bacteria can be utilized to solve the issue of environmental degradation post an oil spill have been attempted [30].

3.3 Mercury entry into the ecosystem

At the human-animal-environment interface, mercury can exist in three forms; elemental (or metallic) mercury, inorganic mercury compounds, and organic mercury compounds. At this point, mercury can be highly persistent, bioaccumulative, and toxic. Mercury occurs naturally in the environment, but it is generally safely contained in minerals and does not present any significant risk, except when anthropogenic activities precipitate its release in large amounts into the environment, consequently circulating freely for a long time [31]. An environment can grossly become polluted with mercury following oil and gas leaks, alkali and metal processing, coal incineration, gold and mercury mining, and from improper medical and other waste disposals. In the environment, microbial organisms can uptake the elemental form of mercury and this signals the transcription of the genes hgcA and hgcB are transcribed to synthesize the HgcA and HgcB proteins [32].

These proteins can then start the methylation reaction to form methylmercury. Mercury and methylmercury exposure to sunlight (specifically ultra-violet light) has an overall detoxifying effect. Sunlight can break down methylmercury to Hg(II) or Hg(0), which can leave the aquatic environment and reenter the atmosphere as a gas as shown in **Figure 4**.

In the food chain, each rung of the food chain consumes more mercury because animals acquire it faster than they expel it. Small ambient quantities of methylmercury can easily build in fish, fish-eating species, and people. Even at low atmospheric deposition rates in remote regions, mercury biomagnification can be harmful to aquatic food chain consumers.

Mercury-associated risks at the wildlife-domestic animal-environment interface.

mercury in water and sediments is the primary concern, as it is in a highly toxic form and can easily be taken up by animals, thus finding its way into the human food chain. Health concerns in Uganda center on human and animal consumption of fish and fish products contaminated with methylmercury. Particularly in humans, neurotoxicity is the most important concern associated with mercury exposure. When methylmercury reaches the bloodstream, it is distributed to all tissues and can cross the normally protective blood–brain barrier to the brain. Methylmercury can also readily move through the placenta to the developing fetus(es) and, therefore, of

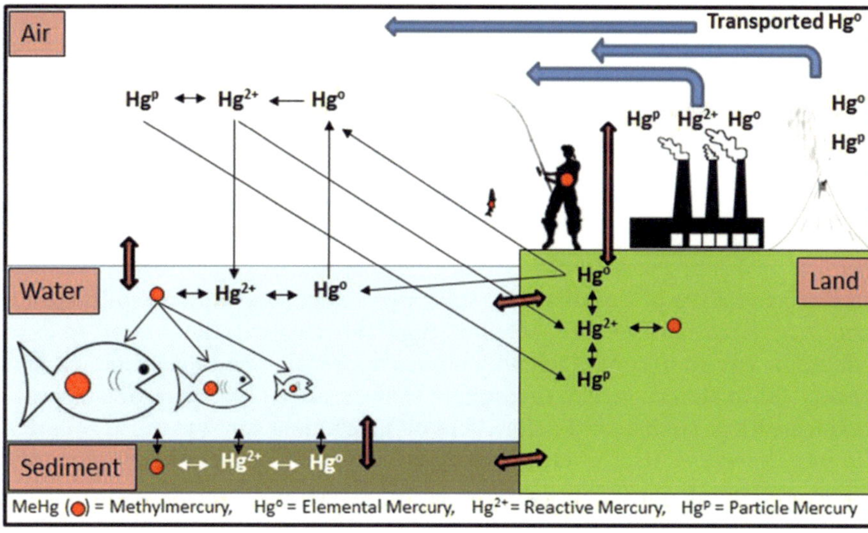

Figure 4.
Mercury's entry into the environment [33].

more concern to pregnant women and women of childbearing age. Mercury exposure in humans (RfD between 1.0 and 3.0×10^{-4} mg/kg per day) can lead to a variety of negative health effects, including neurological, kidney, gastrointestinal, genetic, cardiovascular, and developmental disorders, and even death [32]. Wildlife and domestic animals that consume fish and fish products, depending on their feeding behavior, may ingest large amounts of methylmercury in their diet, often interfering with their reproductive potential. Raptorial bird species, otters, and others that commonly consume water-based foods are at the greater risk of mercury contamination. In birds, reproductive problems are the primary concern related to mercury poisoning. Other mercury effects in wildlife and other animals are liver damage, kidney damage, and neurobehavioral effects.

4. Health risk and burden due to mercury

The exposure diet intake is linked to the Hazard Quotient (HQ), which signifies the relationship between the exposure obtained in the diet and the oral reference dose for mercury [28]. Choice of the oral reference dose is critical in determining the health risks the vulnerable community is exposed to, that is, use of the general population or vulnerable population oral reference dose [34]. The oral reference doses for mercury in the vulnerable and general population are 1×10^{-4} and 3×10^{-4} mg/kg per day, respectively [35, 36]. Therefore, there is a need to send out a message to this vulnerable group about the health hazard they are encountering daily by consumption of food contaminated with mercury. Use of specific messages for different target groups was demonstrated in the USA during a study by Ref. [34], which involved pregnant women and children and observed that there is a need for a unique message for the vulnerable group.

Hazard Index (HI) for both vulnerable and general populations if computed can spell out the health risk. For the vulnerable populations, if the hazard quotients from the study are added and the hazard index is greater than one spells out probable health risks from the mercury consumed. These results are in agreement with Poulin et al. [29] who documented higher HI levels in carnivorous than herbivorous fish, a pointer toward the hazard index points toward noncarcinogenic risk attributed to mercury uptake in fish parts, especially the Lake Albert Nile perch.

5. Conclusions

Mercury and other potentially toxic substances associated with oil spills are a common occurrence in areas of oil exploitation. The negative effects of oil spills can be reduced by environmentally friendly human approaches and the resilience of the ecosystem. Oil exploitation is an investment that should be executed carefully in accordance with guidelines and policy. This oil investment can transform community and country livelihoods positively if managed in accordance with the eco-friendly guidelines and negatively if the sole focus of oil extraction is that of resource generation for country expenditures.

Acknowledgements

We acknowledge the support from Makerere University. We recognize all staff of various colleges who read, advised, and critiqued this work.

I am thankful to the Wildlife and Aquatic Animal Resources Department for spear leading this research. I am grateful to Biosecurity, Ecosystem Health, and Veterinary Public Health Department for the support and guidance, especially Dr. Innocent Rwego.

Last but not the least; I am grateful to the mentors Professor Ejobi Francis, Professor Charles Muyanja, Dr. Irene Naigaga, and Dr. Jesca Nakavuma.

Conflict of interest

There is no known conflict of interest to declare on behalf of the authors.

Notes/thanks/other declarations

Thanks to Intech open Publishers for sharing an opportunity for the publication of the chapter.

Author details

Andrew Tamale[1*], Justine Okello[2] and Celsus Sente[1]

1 COVAB, WAA, Makerere University, Kampala, Uganda

2 COVAB, BEP, Makerere University, Kampala, Uganda

*Address all correspondence to: andietam@gmail.com

References

[1] Chowdhary P, Bharagava RN, Mishra S, Khan N. Role of industries in water scarcity and its adverse effects on environment and human health. In: Environmental Concerns and Sustainable Development. Singapore: Springer; 2020. pp. 235-256 [Accessed: 02 Sep 2022]

[2] Kassim W. 2020. Land conservation in the albertine graben region of uganda: A critical analysis of the legal regimes. In: Springer. pp. 79-99. [Online] Available from: https://link.springer.com/chapter/10.1007/978-3-030-36004-7_5. DOI: 10.1007/978-3-030-36004-7_5 [Accessed: May 19, 2022]

[3] Osuagwu ES, Olaifa E. Effects of oil spills on fish production in the Niger Delta. PLoS One. 2018;**13**(10):e0205114. DOI: 10.1371/JOURNAL.PONE.0205114

[4] Brudler S, Rygaard M, Arnbjerg-Nielsen K, Hauschild MZ, Ammitsøe C, Vezzaro L. Pollution levels of stormwater discharges and resulting environmental impacts. Science of the Total Environment. 2019;**663**:754-763

[5] Cao Y, Li X. Adsorption of graphene for the removal of inorganic pollutants in water purification: A review. Adsorption. 2014;**20**(5):713-727

[6] Balali-Mood M, Naseri K, Tahergorabi Z, Khazdair MR, Sadeghi M. Toxic mechanisms of five heavy metals: Mercury, lead, chromium, cadmium, and arsenic. Frontiers in Pharmacology. 2021;**12**:227. DOI: 10.3389/FPHAR.2021.643972/XML/NLM

[7] Men C, Liu R, Wang Q, Guo L, Shen Z. The impact of seasonal varied human activity on characteristics and sources of heavy metals in metropolitan road dusts. Science of the total Environment. 2018;**637**:844-854

[8] Gao Y et al. Human activities aggravate nitrogen-deposition pollution to inland water over China. National Science Review. 2020;**7**(2):430-440

[9] Sseremba D. Oil and community development in uganda citizens' expectations and participation in the oil and natural gas sector. 2020. [Online] Available from: https://www.theseus.fi/handle/10024/332985. [Accessed: May 19, 2022]

[10] America's Dangerous Pipelines. Available from: https://www.biologicaldiversity.org/campaigns/americas_dangerous_pipelines/. [Accessed Sep. 02, 2022]

[11] Sabela-Rikhotso PTZ, van Niekerk D, Nemakonde LD. Enhancing Coordination for Effective Management of Oil Spill Pollution in South Africa. International Journal of Disaster Risk Science. 2022;**13**:12-24. Available from: www.ijdrs.com. www.springer.com/13753. DOI: 10.1007/s13753-022-00392-8

[12] Molyneaux L, Brown C, Wagner L, Foster J. Measuring resilience in energy systems: Insights from a range of disciplines. Renewable and Sustainable Energy Reviews. 2016;**59**:1068-1079. DOI: 10.1016/J.RSER.2016.01.063

[13] Kabenge I, Katimbo A, Kiggundu N, Banadda N. Bioremediation technology potential for management of soil and water pollution from anticipated rapid industrialization and planned oil and gas sector in Uganda: A review. Journal of Environmental Protection. 2017;**8**(11):1393-1423

[14] Mawejje J. The oil discovery in Uganda's Albertine region: Local expectations, involvement, and impacts.

The Extractive Industries and Society. 2019;**6**(1):129-135

[15] Nnakayima D. Impacts of upstream oil and gas activities on environment, well-being and tourism in the Albertine Region of Uganda: Local community perspectives. Unpublished masters dissertation. Kampala, Uganda: Makerere University. 2018

[16] Qiao K et al. Application of magnetic adsorbents based on iron oxide nanoparticles for oil spill remediation: A review. Journal of the Taiwan Institute of Chemical Engineers. 2019;**97**:227-236

[17] The-ocean-off-the-coast-of-Lima-shines-with-oil-spilled-from-the-refinery-of-La-Pampilla-Peru.webp (WEBP Image, 1536 × 800 pixels) — Scaled (76%). Available from: https://imgs.mongabay.com/wp-content/uploads/sites/20/2022/03/25140533/The-ocean-off-the-coast-of-Lima-shines-with-oil-spilled-from-the-refinery-of-La-Pampilla-Peru.jpg. [Accessed Aug. 30, 2022]

[18] Shell oil spills in the Niger delta - in pictures - Environment - The Guardian. Available from: https://www.theguardian.com/environment/gallery/2011/aug/03/shell-oil-spills-niger-delta-in-pictures. [Accessed Aug. 30, 2022]

[19] Andrew T, Francis E, Charles M, Naigaga I, Jessica N, Micheal O, et al. Mercury concentration in muscle, bellyfat and liver from Oreochromis niloticus and Lates niloticus consumed in Lake Albert fishing communities in Uganda. Cogent Food Agriculture. 31 Dec 2016;**2**(1):1214996. DOI: 10.1080/23311932.2016.1214996. Epub: 2016 Jul 22. PMID: 27774497; PMCID: PMC5059776

[20] Steinhauer I, van Bodegom A, van Dessel B. Review of the Environmental and Social Impact Assessment (ESIA) Report for the Kingfisher Project in Uganda oil spill. 2019. [Online]. Available: https://dspace.library.uu.nl/bitstream/handle/1874/396567/NCEA_7308_NCEA_Review_of_ESIA_Report_for_Kingfisher_Project_Uganda.pdf?sequence=1. [Accessed: Sep. 02, 2022]

[21] Pullin AS et al. Human well-being impacts of terrestrial protected areas. Environmental Evidence. 2013;**2**(1):1-41. DOI: 10.1186/2047-2382-2-19/TABLES/5

[22] Sylvia K. Analysing the effects of oil and gas exploration activities on people's livelihoods in the Albertine graben in western Uganda. 2021. [Online]. Available from: https://kyuspace.kyu.ac.ug/handle/20.500.12504/393. [Accessed: May 19, 2022]

[23] Mitra S, et al. Impact of heavy metals on the environment and human health: Novel therapeutic insights to counter the toxicity. Journal of King Saud University - Science. 2022;**34**(3). DOI: 10.1016/j.jksus.2022.101865

[24] Fuda RK, Ryan SJ, Cohen JB, Hartter J, Frair JL. Assessing impacts to primary productivity at the park edge in Murchison Falls Conservation Area, Uganda. Ecosphere 7.RemoteXs. Wiley Online Libr; 2018;(10):e01486. DOI: 10.1002/ecs2.1486. [Online]. Available from: https://onlinelibrary.wiley.com/doi/abs/10.1111/aje.12568 [Accessed: May 19, 2022]

[25] Liang L, Xi F, Tan W, Meng X, Hu B, Wang X. Review of organic and inorganic pollutants removal by biochar and biochar-based composites. Biochar. 2021;**33**:255-281. DOI: 10.1007/S42773-021-00101-6

[26] Karunasagar I, Girisha SK, Venugopal MN, Biswajit M. Bacteriophage application as a management strategy in shrimp hatcheries. Communications in Agricultural and Applied Biological

Sciences. 2013;**78**(4):204-205 [Online]. Available from: http://www.ncbi.nlm.nih.gov/pubmed/25141668

[27] Tamale A, Ejobi F, Muyanja C, Naigaga I, Nakavuma J, Drago CK, et al. Sociocultural factors associated with fish consumption in Lake Albert fishing community: Guidelines for lead and mercury. Cogent Environmental Science. 21 Mar 2017;**3**(1):1304604. DOI: 10.1080/23311843.2017.1304604. PMID: 30854410; PMCID: PMC6390268

[28] Castilhos ZC et al. Mercury contamination in fish from gold mining areas in Indonesia and human health risk assessment. Science of the Total Environment. 2006;**368**(1):320-325. DOI: 10.1016/J.SCITOTENV.2006.01.039

[29] Poulin J, Gibb H, Prüss-Üstün A, World Health Organization. Mercury: assessing the environmental burden of disease at national and local levels. Geneva: World Health Organization; 2008. Available from: https://apps.who.int/iris/ handle/10665/43875. [Accessed: Sep. 02, 2022]

[30] Ssenku JE, Walusansa A, Oryem-Origa H, et al. Bacterial community and chemical profiles of oil-polluted sites in selected cities of Uganda: Potential for developing a bacterial-based product for remediation of oil-polluted sites. BMC Microorganisms. 2022;**22**:120. DOI: 10.1186/s12866-022-02541-x

[31] Rice KM, Walker EM, Wu M, Gillette C, Blough ER. Environmental mercury and its toxic effects. Journal of Preventive Medicine and Public Health. 2014;**47**(2):74-83. DOI: 10.3961/JPMPH.2014.47.2.74

[32] Tang WL, Liu YR, Guan WY, Zhong H, Qu XM, Zhang T. Understanding mercury methylation in the changing environment: Recent advances in assessing microbial methylators and mercury bioavailability. Science of the Total Environment. 20 Apr 2020;**714**:136827. DOI: 10.1016/j.scitotenv.2020.136827. Epub: 2020 Jan 22. PMID: 32018974

[33] Floreani F, Acquavita A, Barago N, Klun K, Faganeli J, Covelli S. Gaseous mercury exchange from water-air interface in differently impacted freshwater environments. International Journal of Environmental Research and Public Health. 2022;**19**(13):8149. DOI: 10.3390/ijerph19138149

[34] What to tell your clients about eating fish. Available from: https://agris.fao.org/agris-search/search.do?recordID=US201300990192. [Accessed Sep. 03, 2022]

[35] Zhu L, Yan B, Wang L, Pan X. Mercury concentration in the muscle of seven fish species from Chagan Lake, Northeast China. Environmental Monitoring and Assessment. 2012;**184**(3):1299-1310. DOI: 10.1007/S10661-011-2041-7

[36] Andrew T, Francis E, Charles M, Irene N, Jesca N, Ocaido M, et al. Risk estimates for children and pregnant women exposed to mercury-contaminated Oreochromis niloticus and Lates niloticus in Lake Albert Uganda. Cogent Food Agriculture. 31 Dec 2016;**2**(1):1228732. DOI: 10.1080/23311932.2016.1228732. Epub: 2016 Sep 8. PMID: 27722183; PMCID: PMC5039401

Chapter 10

Mercury Pollution: Dangers and Treatment

Fattima Al-Zahra Gabar Gassim

Abstract

Mercury (Hg) is a toxic heavy metal with interesting properties such as silvery-white liquid at room temperatures, volatile, a poor conductor of heat, but a fair conductor of electricity. Mercury contamination in soil, water, and the air are associated with potential toxicity to humans and ecosystems. The nervous system is very sensitive to all forms of mercury. Exposure to high levels of any type of mercury can permanently damage the brain, kidneys, and developing fetus. Mercury can build up in the bodies of fish in the form of methyl mercury(organic mercury) which is very poisoning and largely linked to eating seafood, mainly fish. The mechanism of the mercury poisoning treatment involves adsorption, oxidation, and reduction processes. The major aim of these technologies is to separate mercury from the contaminated media or transform toxic mercury species into less toxic ones.

Keywords: mercury sources, environmental pollutions, mercury treatments, heavy metals, mercury dangerous

1. Introduction

Mercury (Hg) is an element like lead or arsenic, and it is classified as heavy metal. It is the only metal present in liquid form as shown in **Figure 1**. It has many interesting properties. The most important properties was listed in **Table 1**.

Mercury is a naturally occurring mineral found primarily in a mineral called cinnabar, which can contain up to 86 percent mercury. It is released through natural rock weathering and (or) volcanic activity [2].

Metallic mercury is used in the production of chlorine gas and caustic soda, and it is also used in thermometers, dental fillings, and batteries. Mercury salts are sometimes used in skin-lightening creams and as antiseptic creams and ointments [3].

1.1 Facts about mercury

a. Highly toxic to the nervous system and damages memory, cognitive thinking, language abilities, attention, and fine motor skills [4].

b. Mercury (Hg) is ubiquitous, naturally enriched in volcanic regions

Figure 1.
Drops stock of mercury.

Values	Properties
Molecular formula	Hg
atomic number	80
Atomic weight	200.592 amu
Valence	1, 2
Melting point	−39°C
Boiling point	357°C
Specific gravity	13.6
Vapors pressure	0.0012 (mm Hg/21°C)
Odor	Odorless
Stability	Stable and does not tarnish and is slightly volatile at ordinary temperatures
Solubility	Not soluble in water or most other liquids, but will dissolve in lipids (fats and oils)
Conductivity	An excellent conductor of electricity

Table 1.
Some of the important properties of Hg [1].

c. Bio-accumulative (higher concentrations in tissues of aquatic plants and animals than in water) [5].

d. Biomagnified (higher concentrations at increasingly higher levels in the food chain) [6].

e. Many chemical forms in the air, water, sediment, and living organisms

1.2 Sources of mercury

Natural levels of mercury exist in soil, air, and water around the world [7]. Mercury is introduced into the environment in three ways.

1. Mercury is naturally emitted into the air by volcanoes, rocks, and forests fires and soil. Mercury is released into the atmosphere due to its evaporation, it can easily move through the air and end up thousands of kilometers from where it was first released, and traveled hundreds of miles with the wind before being deposited on the surface of the earth [8]. Sedimentation can occur as little as five to fourteen days afterward Mercury is released into the air, or it can take up to one year - during which time the mercury can they reside in the air and are transported to distant places around the world. Once deposited on Earth, mercury can onto the surface waters of the state [9].

2. Mercury is emitted into the air from the combustion of fossils Fuel and municipal or medical waste. Mercury can enter the environment through human activities such as burning coal, extracting minerals from ore, manufacturing cement, and using and disposing of mercury-containing products, such as fluorescent lamps and some types of batteries [10]. In certain regions of the world, small-scale gold mining with mercury is an important source of mercury pollution. Human emissions account for 40% of the mercury deposited in Canada each year, and 97% of these emissions come from other countries [11].

3. Mercury can be reintroduced into the environment through natural processes such as the evaporation of ocean waters. Mercury persists in the environment for long periods by circulating back and forth between air and soil [12], every time chemical shapes change. Atmospheric ages can be oxidized and combined with other elements, such as chlorine, sulfur, or oxygen, to form inorganic mercury)compounds (HgS, $HgCl$, HgO). These inorganic compounds are estimated to be up to 2 years,

Also, elemental mercury can be combined with carbon to form more toxic organic methyl mercury(CH_3Hg) which remains in the soil for decades. Mercury is never removed from the environment; It was moved to other sites and eventually buried under soil and sediment. **Figure 2** [13] shows the Inter-phase transfer and transport of mercury in soil, water, and air.

1.3 Dangers of mercury exposure

Three different types of mercury are harmful to the human body:

- Elemental mercury (liquid mercury, mercury silver): They are found in fluorescent lights, switches, glass thermometers, and dental fillings.

- Inorganic mercury: It is found in chemistry laboratories, certain types of disinfectants, and in batteries.

- Organic mercury: It is found in the fumes of coal and fish that have ingested methylmercury and disinfectants (germ killers such as red mercury) [14].

The human body can be exposed to mercury through [15]:

- The skin by touching it
- Air by inhalation
- Eating or drinking contaminated food or water

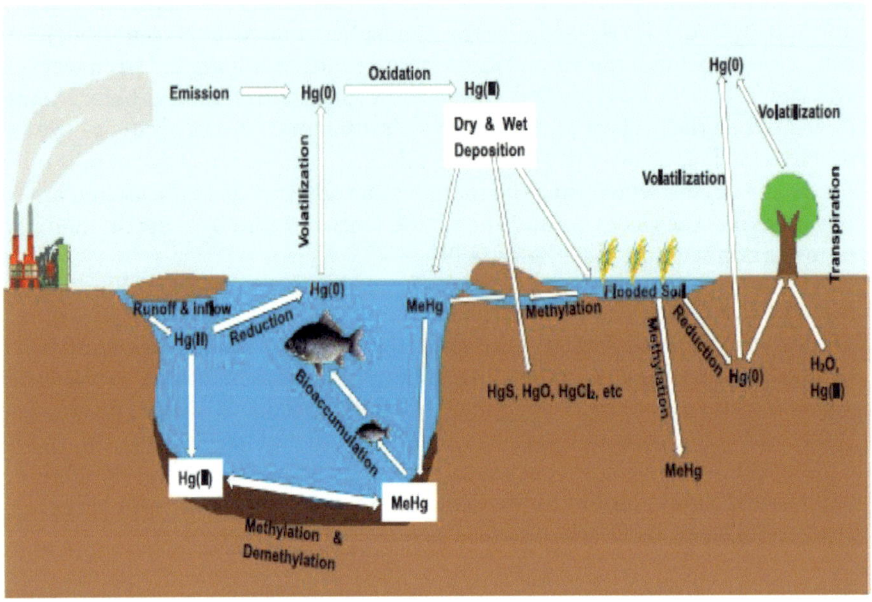

Figure 2.
Inter-phase transfer and transport of mercury in soil, water, and air. Acronyms: Hg(0), elemental mercury; Hg(II), divalent mercury; MeHg, methyl mercury.

The nervous system is very sensitive to all forms of mercury. Exposure to high levels of any type of mercury can permanently damage the brain, kidneys, and fetus. The effects on brain function may cause irritability, shyness, tremors, changes in vision, or problems with hearing and memory [16]. High exposure to mercury vapor can cause chest pain, shortness of breath, and fluid buildup in the lungs (pulmonary edema) that can be fatal [17].

Elemental mercury can turn into more toxic inorganic compounds as oxidized mercury (Hg^{2+}) combines with other elements, or it can combine with carbon to form an even worse pollutant known as methyl mercury (CH_3Hg). These compounds may fall to land or water through precipitation, or they may fall as dry particles and find their way into a lake or ocean [4].

Methyl mercury and metallic mercury fumes are particularly harmful because more mercury reaches the brain. Long-term exposure may cause blurring of the eye. Contact with mercury chloride can cause skin burns and permanent eye damage. Mercury also accumulates in the body [18]. Most metallic mercury will accumulate in the kidneys, but some metallic mercury can also accumulate in the brain. Most of the metallic mercury that is absorbed by the body is eventually left in the urine and feces, while small amounts leave the body in the same exhalation [19].

Humans are exposed to mercury in two ways:

1. Eating fish contaminated with organic methyl mercury

Symptoms of organic mercury poisoning from long-term exposure include a feeling of numbness or pain in certain parts of the body, tremors (uncontrollable shaking), unsteady walking, double vision, or blurry vision. Blindness, memory loss, and seizures [20].

Mercury can enter the open seas and oceans as a result of downstream movement and re-deposition of polluting sediments from urban estuaries. The reduction and oxidation of mercury mostly occur near the surface of ocean waters. These are either driven by sunlight or microbial activity. Under ultraviolet rays, elemental mercury oxidizes and dissolves directly in ocean waters or binds to other molecules [21]. The reverse reaction reduces some Hg^{2+} to elemental Hg^0 and returns to the atmosphere. Atmospheric fine aerosols such as ocean water droplets can act as small reaction chambers in this process providing the required special reaction conditions. Oxidation and reduction of mercury in the ocean are not very simple reversible reactions. The proposed photochemical pathway for mercury surrounding ocean aerosols was shown in **Figure 3** which indicates that it occurs through a reactive medium [22].

Photo-oxidation is suspected to be OH-driven. Roots and reduction are driven by wind and perturbations of the surface layer.

In the dark, mercury redox reactions continue due to microbial activity. Biological transformations vary and have a lower rate compared to the above sunlight-driven processes. Inorganic mercury Hg^{2+} and methyl mercury can adsorb in molecules. A positive correlation was observed between the amount of organic matter versus the concentration of these types of mercury, indicating that most of them are associated with organic matter [23]. This phenomenon can determine the bioavailability and toxicity of mercury in the ocean. Some methyl mercury is released into the ocean through river run-off. However, most of the methyl mercury found in the ocean is produced in ceto (within the ocean itself). Inorganic Hg methylation can occur via biotic and abiotic pathways. However, biosynthetic pathways are the most prevalent. The reactions shown in the simplified diagram below are parts of the complex enzyme-driven metabolic pathways that occur within microbial cells [24].

In abiotic reactions, hemic substances act as methylating agents, and thus this process occurs at shallow sea levels where decomposing organic matter is available to combine with inorganic Hg^{2+}. Mercury methylation studies in polar regions have also shown a positive correlation between methylation and chlorophyll content in water. The potential for biological pathways to produce methyl mercury is

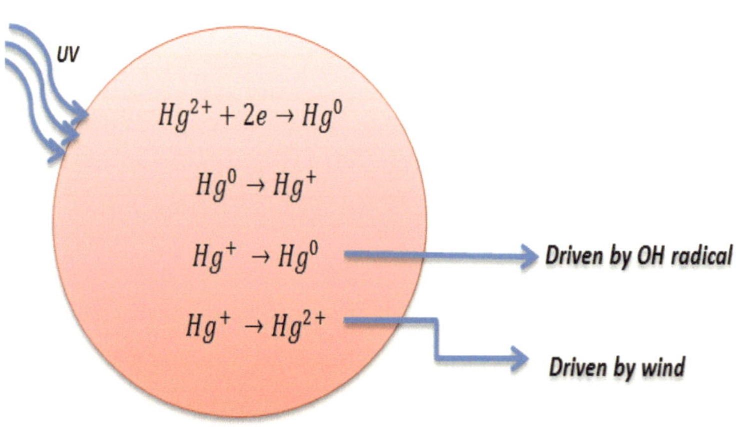

Figure 3.
Photochemistry of mercury on oceanic aerosols.

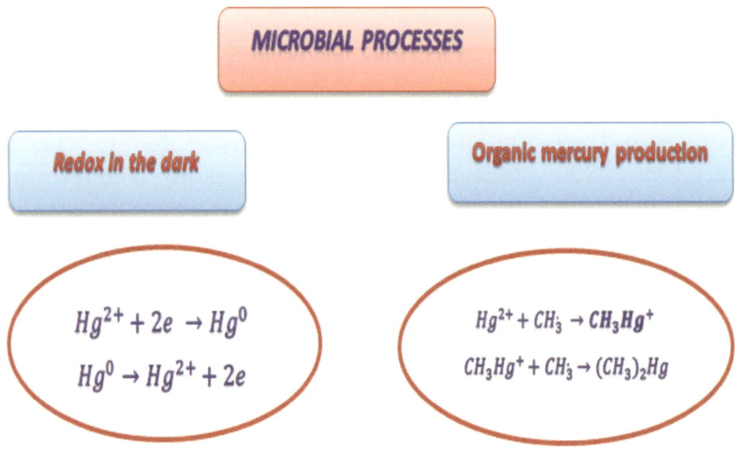

Figure 4.
Microbial chemical conversions of mercury.

shown in **Figure 4** [25]. The methyl mercury produced accumulates in microbes. Due to the high permeability and absence of methyl mercury degradation in other species that depend on those microbes, this highly biotoxic compound is amplified through marine food chains to top predators. Humans consume many types of marine fish that are the number one predators in food chains, putting their health at great risk. Therefore, finding possible solutions to further reduce mercury emissions and clean up existing mercury emissions and clean up existing mercury pollution [26].

Methyl mercury is toxic and can cause very harmful effects when consumed, which can happen when humans eat highly contaminated fish. Through the process of biomagnification, as shown in **Figure 5**, the concentration of methyl mercury within fish increases as one goes up the food chain. This makes eating apex predators dangerous, and especially dangerous for pregnant women and young children to do so.

Mercury levels in fish are measured in either part per million (mg kg^{-1}) or dry weight micrograms (mcg). **Table 2** explains some common types of tuna and the concentrations of mercury in them [27].

The effects of methyl mercury can lead to risk neurological problems, especially in young children and infants, by affecting the brain and nervous system. Possible problems include cerebral palsy, delayed walking or speech, learning difficulties, tremors, irritability, poor coordination, and memory loss. Pregnant mothers in particular should not eat large fish because their babies are susceptible to these chemicals that attack developing organs [28].

The US Environmental Protection Agency (EPA) states that 0.045 micrograms of mercury per pound (0.1 micrograms per kilogram) of body weight per day are the maximum safe dose of mercury. This amount is known as the reference dose [29].

The daily reference dose of mercury is based on body weight. Multiplying that number by seven gives you the weekly mercury limit. **Table 3** shows some examples of reference doses based on different body weights [30].

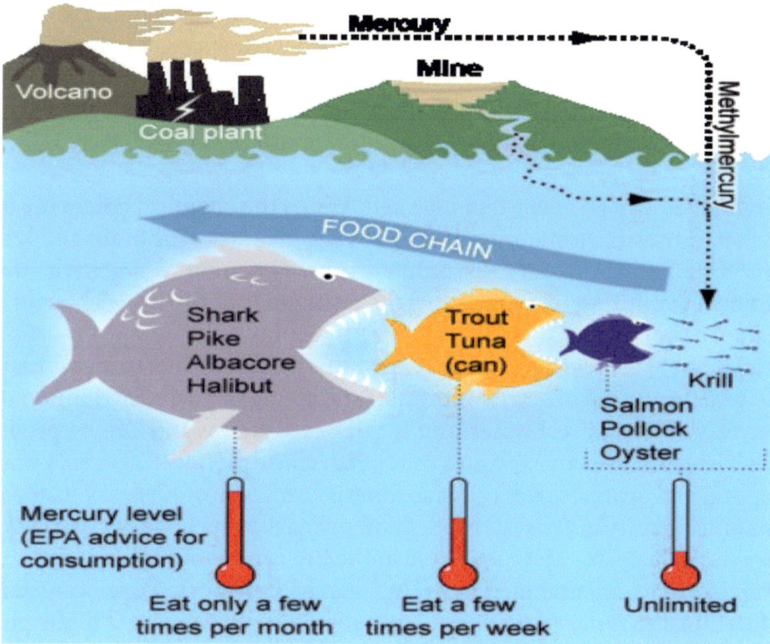

Figure 5.
Methyl mercury concentrations within the fishes.

Species	Mercury in mg kg^{-1}	Mercury (in mcg) per 3 ounces (85 grams)
Light tuna (canned)	0.126	10.71
Skipjack tuna (fresh or frozen)	0.144	12.24
Albacore tuna (canned)	0.350	29.75
Yellow fin tuna (fresh or frozen)	0.354	30.09
Albacore tuna (fresh or frozen)	0.358	30.43
Big eye tuna (fresh or frozen)	0.689	58.57

Table 2.
Some common types of tuna and the concentrations of mercury in them.

Body weight	Reference dose per day (in mcg)	Reference dose per week (in mcg)
100 pounds (45 kg)	4.5	31.5
125 pounds (57 kg)	5.7	39.9
150 pounds (68 kg)	6.8	47.6
175 pounds (80 kg)	8.0	56.0
200 pounds (91 kg)	9.1	63.7

Table 3.
Some examples of reference doses based on different body weights.

Since some tuna species are very high in mercury, a single 3-ounce (85-gram) serving may have a mercury concentration that equals or exceeds a person's weekly reference dose.

2. Inhalation of elemental mercury (Hg) or inorganic salts (Hg^{2+})

Elemental mercury is toxic when ingested. When the chemical enters the body by inhalation, it travels through the bloodstream and attacks the brain and kidneys. Symptoms of inorganic mercury poisoning involve a burning feeling in the throat and/or stomach, vomiting or nausea, diarrhea, the color of the urine changes and blood in stool or vomit [20].

The atmosphere is the primary pathway for mercury transport emissions, while land and ocean processes play an important role Its role in the redistribution of mercury in terrestrial water, freshwater, and Marine ecosystems and CH_3Hg production that drives The main human exposure route, fish consumption, Especially marine fish. Temporal and spatial scales of Atmospheric transport of mercury to aquatic organisms and Terrestrial ecosystems depend primarily on chemicals and physical forms. **Figure 6** illustrates Mercury Cycle in the Environment [31].

After emission, elemental mercury (Hg^0) can be transported over long distances before oxidation and removal by dry precipitation of particles or gas in the phase or cleaning by scavenging precipitation. The atmospheric residence time of Hg^0 is from several months to years and therefore mercury can be transported and deposited in remote locations such as the Arctic and Antarctic [32].

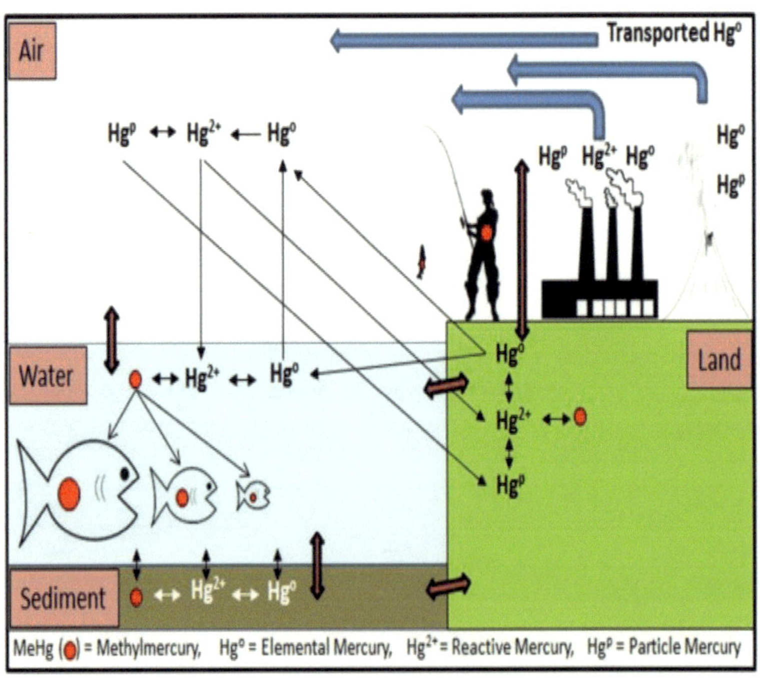

Figure 6.
Mercury cycle in the environment.

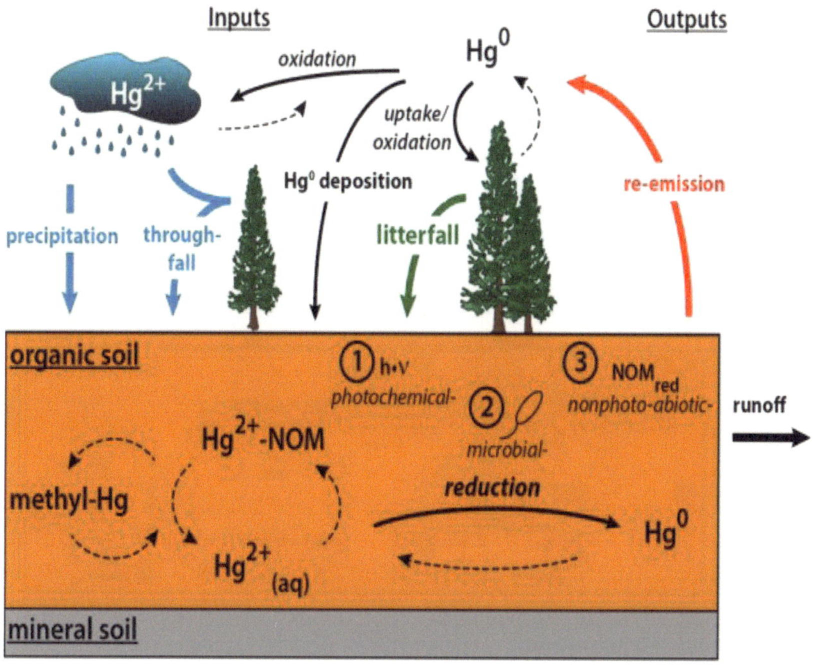

Figure 7.
Current estimates of the fluxes and pools of mercury at the Earth's surface.

The ionic particle bound Hg^{2+} has a shorter Atmospheric residence time of Hg^0 (atmospheric residence hours to days), as a result of which it is generally deposited locally or regionally. Inputs to ecosystems occur to a large extent Hg^{2+}, while most CH_3Hg is produced within ecosystems. It is important to distinguish between primary and secondary mercury emission sources. Primary sources, both natural and Man-made mercury from the long-lived lithosphere atmospheric reservoirs. This mercury is deposited in the earth and oceans. Precipitated mercury can be reduced to Hg^0 and then re-emitted [33].

Re-emissions are secondary sources of the exchange of mercury between surface reservoirs using the atmosphere car. Primary sources increase the global pool of mercury in Surface reservoirs, while secondary sources redistribute Between and within ecosystems [12].

Mercury is deposited into the atmosphere primarily as oxidized mercury (II).

By precipitation and falling or picked up by plant stomata and deposited with excreta. In soil, Hg(II) can be reduced by different pathways:

(i) Photochemical, (ii) microbial, or (iii) abiotic non-chemical reduction by natural organic matter (NOM), followed by re-emission back into the atmosphere. All forms of mercury are subject to leaching from soils with runoff or groundwater into aquatic ecosystems. **Figure 7** shows emission and re-emission sources of Hg^0 [34].

2. Mercury poisoning treatment

As known, even a small amount of mercury can affect the digestive, nervous, and immune systems. Also, it can be a threat to the development and growth of a child in early life. Mercury products are hazardous waste. When this waste is placed in the trash, it does not decompose. Instead, they find their way into lakes, rivers, or soil [35].

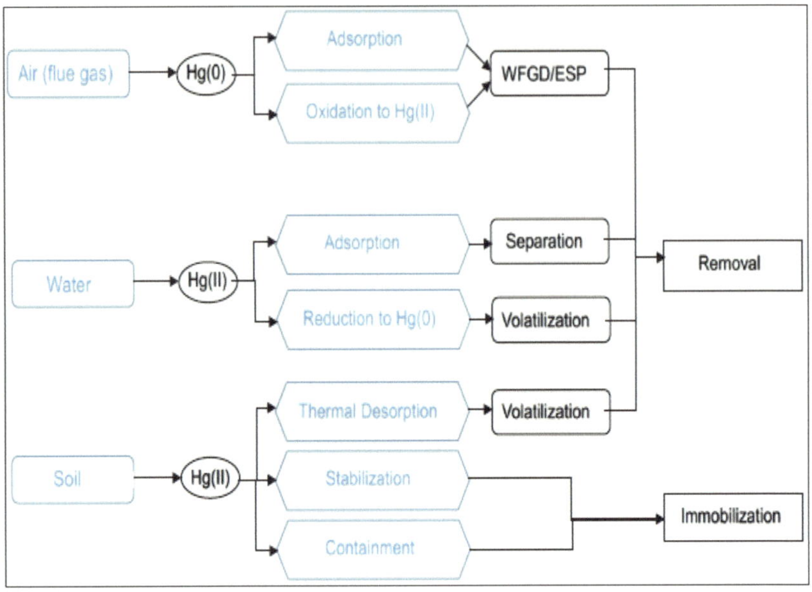

Figure 8.
Major mechanisms involved in Hg treatment. Acronyms: WFGD, wet flue gas desulfurization; ESP, electrostatic precipitation.

Mercury like other heavy metals, cannot be degraded in ecosystems, thus treatment should be based on removals or installations. Removal techniques include adsorption mechanism, Adsorption, oxidation, and reduction [36]. The main objective of these techniques is the separation or conversion of mercury from contaminated media toxic mercury species to less toxic species. The most widely adopted installation techniques are stabilization and containment, Which prevents the transfer of mercury by chemical complexity or physical baiting, respectively [37].

Novel materials, especially materials that possess high Surface area, large porosity as well as adsorption active sites It has been extensively examined in recent studies Regardless of the absorption capacity which is the main determinant of these materials, and other issues such as the method of generation, Stability, and reusability should also be seriously considered compared to traditional processing techniques such as heat absorbent or activated carbon adsorption, innovative [38].

The methods have proven to be more cost-effective and environmentally friendly. Interestingly, most of these techniques treat contaminated mercury Soil, water, and air can depend on the emerging materials or metabolizing organisms, i.e. plants, algae, and bacteria. **Figure 8** [15] shows the main mechanisms involved in Hg treatment.

Major mechanisms involved in Hg treatment from the air, water, and soil can be explained as follows:

a. The catalyst oxidation method is commonly used for gaseous phase mercury removal. Oxidation is a more cost-effective method to remove Hg^0 from the flue gas. Several studies have focused on new oxidation, regardless of the oxidation method of the catalyst, the advanced free radical oxidation of Hg^0 is also sometimes used, but the removal capacity of Hg^0 using this technology is still limited [39].

The synthesis of a functional covalent thioether of triazine nanoparticles for Hg^{2+} and Hg^0 is absorbed by water studies and the results show an excellent adsorption capacity (1253 and 813 mg/g for mercury Hg^{2+} and Hg^0, respectively). The maximum adsorption capacity was reached 172.6 mg/g by using silica-coated magnetron nanoparticles for Hg(II) extraction from wastewater and adsorption of mercury ions on imine groups (C-NH-) on the surface of nanoparticles [40].

b. Reduction of Hg^{2+} to Hg^0 is often applied to prevent the formation of Methyl mercury (MeHg). Methyl mercury is the most bioavailable form of Mercury. Oxidation-reduction conditions in wetland sediments enhance formation From MeHg. The high concentration of mercury (II) species results in production from MeHg. Therefore, an effective way to control the production of methyl mercury is to reduce the Hg(II) concentration. Zerovalent iron (ZVI) or Fe (II) is often used. To reduce Hg(II) to Hg(0), thus inhibiting the production of MeHg [41].

Adsorption methods by using adsorbents usually possess high surface area and high porosity and the formation of chelates is the major approach to removing Hg(II) from water solution [42].

c. Stabilization approaches freeze the movement of mercury into contaminated sites. Through chemical complexity to reduce solubility for Reduce exposure to mercury in the environment During chemical fixation Sulfur-containing reagents such as elemental sulfur and pyrite (FeS_2) or thiosulfate are commonly used to react with Hg(0) in pollutants oil to form HgS, which is very insoluble [43].

Soil can be dealt with either on-site or off-site, the former requires less energy and labor cost. However, fine mixing is still very difficult on-site stability. The basic defect in stability is that mercury is not removed from contaminated media, Thus it requires permanent prospective monitoring of contaminants on site. Similar to installation, pollution is left on site during containment treatment. Low permeability physical barriers (eg clay plaster walls, coverings, or curtains) around contaminated soil To isolate and contain the soil, thus preventing the migration of mercury to the surrounding environment. These material Barriers can be divided into three types: barricades, vertical barricades, and horizontal barriers [44].

Nano-materials are gaining more and more attention in mercury remediation of soil, water, and flue gas, owing to their high adsorption capacity, small dimension, and another unique electrical, mechanical and chemical properties. There are several major types of nanoparticles and nanocomposites such as carbon nanotubes (CNT) [45], Zinc oxide (ZnO) [46], and Ferro ferric oxide (Fe_3O_4), [47] nanoparticles can be used for Hg remediation in the wastewater.

3. Conclusions

Mercury is one of the most dangerous pollutants which cycles through the atmosphere, water, and soil in various forms to different parts of the world. High levels of Mercury exposure cause damage to the brain, kidneys, and fetus. The effects on brain function may cause irritability, shyness, tremors, changes in vision, or problems with hearing and memory. Humans are exposed to mercury either by Eating fish contaminated with organic methyl mercury or Inhaling elemental mercury (Hg) or inorganic salts (Hg^{2+}). Mercury poisoning treatment techniques include adsorption, oxidation, and reduction mechanism. The catalyst oxidation method is commonly used for gaseous phase mercury removal, while nanoparticles and nanocomposites can be used for mercury remediation of soil, water, and flue gas due to their high adsorption capacity.

Author details

Fattima Al-Zahra Gabar Gassim
College of Pharmacy, University of Babylon, Hilla, Iraq

*Address all correspondence to: alzahraafatema6@gmail.com

References

[1] Ali M, Yaser MM. Structures and properties of mercury (II) coordination polymers. Coordination Chemistry Reviews. 2009;**253**(13-14):1882-1905

[2] Sverdrup HU, Olafsdottir AH. System dynamics modelling of the global extraction, supply, price, reserves, resources and environmental losses of mercury. Water, Air, & Soil Pollution. 2020;**231**(8):1-22

[3] Kefi I, Maria A. Uses and effects of mercury in medicine and dentistry. Journal of Ayub Medical College Abbottabad. 2012;**24**(3-4):204-207

[4] Clarkson TW, Magos L. The toxicology of mercury and its chemical compounds. Critical Reviews in Toxicology. 2006;**36**(8):609-662

[5] Magos L, Clarkson TW. Overview of the clinical toxicity of mercury. Annals of Clinical Biochemistry. 2006;**43**(4):257-268

[6] O'Connor D, Hou D, Ok YS, Mulder J, Duan L, Wu Q, et al. Mercury speciation, transformation, and transportation in soils, atmospheric flux, and implications for risk management: A critical review. Environment International. 2019;**126**:747-761

[7] Chen L, Liang S, Liu M, Yi Y, Mi Z, Zhang Y, et al. Trans-provincial health impacts of atmospheric mercury emissions in China. Nature Communications. 2019;**10**(1):1-2

[8] Selin NE, Wu S, Nam KM, Reilly JM, Paltsev S, Prinn RG, et al. Global health and economic impacts of future ozone pollution. Environmental Research Letters. 2009;**4**(4):044014

[9] Sprovieri F et al. A review of worldwide atmospheric mercury measurements. Atmospheric Chemistry and Physics. 2010;**10**(17):8245-8265

[10] Hopkinson Gordon R, Goodman Teresa M, Prince Stuart R. A Guide to the Use and Calibration of Detector Array Equipment. SPIE Press; 2004

[11] Berthiaume A. Use of the national pollutant release inventory in environmental research: A scoping review. Environmental Reviews. 2021;**29**(3):329-339

[12] Mason RP, Choi AL, Fitzgerald WF, Hammerschmidt CR, Lamborg CH, Soerensen AL, et al. Mercury biogeochemical cycling in the ocean and policy implications. Environmental Research. 2012;**119**:101-117

[13] Nelson Peter F et al. Atmospheric mercury emissions in Australia from anthropogenic, natural and recycled sources. Atmospheric Environment. 2012;**62**:291-302

[14] Mahaffey Kathryn R. Mercury exposure: Medical and public health issues. Transactions of the American Clinical and Climatological Association. 2005;**116**:127

[15] Liuwei W et al. Remediation of mercury contaminated soil, water, and air: A review of emerging materials and innovative technologies. Environment International. 2020;**134**:105281

[16] Lo C, Taylor RS, Gamble G, McCowan L, North RA. Use of automated home blood pressure monitoring in pregnancy: Is it safe? American Journal of Obstetrics and Gynecology. 2002;**187**(5):1321-1328

[17] Harada M. Minamata disease: Methylmercury poisoning in Japan

caused by environmental pollution. Critical Reviews in Toxicology. 1995;**25**(1):1-24

[18] WienerJG, KrabbenhoftDP, HeinzGH, Scheuhammer AM. Ecotoxicology of mercury. In: Handbook of Ecotoxicology. CRC Press; 2002. pp. 433-488

[19] Liang YX, SUN RK, Sun Y, Chen ZQ, Li LH. Psychological effects of low exposure to mercury vapor: Application of a computer-administered neurobehavioral evaluation system. In: Neurobehavioral Methods and Effects in Occupational and Environmental Health. Academic Press; 1994. pp. 133-140

[20] Risk, Who Is At, et al. Agency for Toxic Substances and Disease Registry Case Studies in Environmental Medicine (CSEM) Nitrate/Nitrite Toxicity Where Can I Find More Information?

[21] Kirk JL, Lehnherr I, Andersson M, Braune BM, Chan L, Dastoor AP, et al. Mercury in Arctic marine ecosystems: Sources, pathways and exposure. Environmental Research. 2012:119, 64-187

[22] Qureshi A, O'Driscoll NJ, MacLeod M, Neuhold YM, Hungerbühler K. Photoreactions of mercury in surface ocean water: Gross reaction kinetics and possible pathways. Environmental Science & Technology. 2010;**44**(2):644-649

[23] Beckers F, Rinklebe J. Cycling of mercury in the environment: Sources, fate, and human health implications: A review. Critical Reviews in Environmental Science and Technology. 2017;**47**(9):693-794

[24] Tchounwou PB, Ayensu WK, Ninashvili N, Sutton D. Environmental exposure to mercury and its toxicopathologic implications for public health. Environmental Toxicology: An International Journal. 2003;**18**(3):149-175

[25] Hong YS, Kim YM, Lee KE. Methylmercury exposure and health effects. Journal of Preventive Medicine and Public Health. 2012;**45**(6):353

[26] Outridge PM, Mason RP, Wang F, Guerrero S, Heimburger-Boavida LE. Updated global and oceanic mercury budgets for the United Nations Global Mercury Assessment 2018. Environmental Science & Technology. 2018;**52**(20):11466-11477

[27] Bruce L. Mercury in the Environment. desLibris; 2003

[28] US Department of Health and Human Services. Centers for Disease Control and Prevention Agency for Toxic Substances and Disease Registry. (nd). Your guide to clear writing. Atlanta, GA: ATSDR;

[29] Silbernagel SM, Carpenter DO, Gilbert SG, Gochfeld M, Groth E, Hightower JM, et al. Recognizing and preventing overexposure to methylmercury from fish and seafood consumption: Information for physicians. Journal of Toxicology. 2011

[30] Groth E III. Ranking the contributions of commercial fish and shellfish varieties to mercury exposure in the United States: Implications for risk communication. Environmental Research. 2010;**110**(3):226-236

[31] Sundseth K, Pacyna JM, Pacyna EG, Pirrone N, Thorne RJ. Global sources and pathways of mercury in the context of human health. International Journal of Environmental Research and Public Health. 2017;**14**(1):105

[32] Durnford D, Dastoor A, Figueras-Nieto D, Ryjkov A. Long range transport

of mercury to the Arctic and across Canada. Atmospheric Chemistry and Physics. 2010;**10**(13):6063-6086

[33] Pacyna EG, Pacyna JM, Steenhuisen F, Wilson S. Global anthropogenic mercury emission inventory for 2000. Atmospheric Environment. 2006;**40**(22):4048-4063

[34] Jiskra M, Wiederhold JG, Skyllberg U, Kronberg RM, Hajdas I, Kretzschmar R. Mercury deposition and re-emission pathways in boreal forest soils investigated with Hg isotope signatures. Environmental Science & Technology. 2015;**49**(12):7188-7196

[35] Roman HA, Walsh TL, Coull BA, Dewailly É, Guallar E, Hattis D, et al. Evaluation of the cardiovascular effects of methylmercury exposures: Current evidence supports development of a dose–response function for regulatory benefits analysis. Environmental Health Perspectives. 2011;**119**(5):607-614

[36] Lewis AS, Huntington TG, Marvin-DiPasquale MC, Amirbahman A. Mercury remediation in wetland sediment using zero-valent iron and granular activated carbon. Environmental Pollution. 2016;**212**:366-373

[37] He F, Gao J, Pierce E, Strong PJ, Wang H, Liang L. In situ remediation technologies for mercury-contaminated soil. Environmental Science and Pollution Research. 2015;**22**(11):8124-8147

[38] Gusain R, Kumar N, Fosso-Kankeu E, Ray SS. Efficient removal of Pb (II) and Cd (II) from industrial mine water by a hierarchical MoS2/SH-MWCNT nanocomposite. ACS Omega. 2019;**4**(9):13922-13935

[39] Chen W, Pei Y, Huang W, Qu Z, Hu X, Yan N. Novel effective catalyst for elemental mercury removal from coal-fired flue gas and the mechanism investigation. Environmental Science & Technology. 2016;**50**(5):2564-2572

[40] Abbas K, Znad H, Awual MR. A ligand anchored conjugate adsorbent for effective mercury (II) detection and removal from aqueous media. Chemical Engineering Journal. 2018;**334**:432-443

[41] Amirbahman A, Kent DB, Curtis GP, Marvin-DiPasquale MC. Kinetics of homogeneous and surface-catalyzed mercury (II) reduction by iron (II). Environmental Science & Technology. 2013;**47**(13):7204-7213

[42] AlOmar MK, Alsaadi MA, Hayyan M, Akib S, Ibrahim M, Hashim MA. Allyl triphenyl phosphonium bromide based DES-functionalized carbon nanotubes for the removal of mercury from water. Chemosphere. 2017;**167**:44-52

[43] Piao H, Bishop PL. Stabilization of mercury-containing wastes using sulfide. Environmental Pollution. 2006;**139**(3):498-506

[44] Mulligan CN, Yong RN, Gibbs BF. Remediation technologies for metal-contaminated soils and groundwater: An evaluation. Engineering Geology. 2001;**60**(1-4):193-207

[45] Alimohammady M, Jahangiri M, Kiani F, Tahermansouri H. Design and evaluation of functionalized multi-walled carbon nanotubes by 3-aminopyrazole for the removal of Hg (II) and As (III) ions from aqueous solution. Research on Chemical Intermediates. 2018;**44**(1):69-92

[46] Gabar Gassim FA, Makkaw AJ, Aysa NH. Removal of mercury (II) in aqueous solution by using ZnO and ZnO/CdS nanoparticles as photocatalysts. Iranian Journal of Catalysis. 2021;**11**(4):397-403

[47] Bao S, Li K, Ning P, Peng J, Jin X, Tang L. Highly effective removal of mercury and lead ions from wastewater by mercaptoamine-functionalised silica-coated magnetic nano-adsorbents: Behaviours and mechanisms. Applied Surface Science. 2017 Jan;**30**(393):457-466